Beginner's Guide to 3D Printing

By Chuck Hellebuyck

Published by Electronic Products.

The publisher offers special discounts on bulk orders of this book.

For information contact:

Electronic Products
P.O. Box 251
Milford, MI 48381
www.elproducts.com
chuck@elproducts.com

Printed in United States of America

Table of Contents

Chapter 1 – What is 3D Printing

A three dimensional or 3D object is something we see, touch or use every day. It's an item that has a width, length and height or three dimensions. 3D printing involves building an object in thin layers that are stacked on top of each other to create a 3D solid object. It's like building a cake with many layers as thin as paper. The shape of that wedding cake could be far more complex if the diameter of each layer was slightly different. A cake the shape of a wedding ring could easily be created with rounded edges. 3D printing has been around for many years but only in the last couple of years has it been possible for someone to afford a 3D printer at home. This allows anyone to create his or her own 3D printed designs.

3D prints are created in software by first taking a solid object and then slicing it into many very thin layers. A file with each layer's x, y and z dimensions are sent to a 3D printer. Each layer is printed with a material that fuses to the previous layer. The typical home 3D printer uses melted plastic as the material to form the final solid object.

3D printing can be accomplished in many different ways. Some 3D printers use a laser to harden a special plastic liquid or harden a special plastic powder. As each layer is hardened, a base pulls the hardened layer down a slight bit and then the liquid or powder fills in on top of it and then the laser hardens the next layer right on top of the previous layer. Some 3D printers extrude the plastic directly and just lay down a softened layer of plastic on a base like a thin line of toothpaste. The printer then lowers the base so a new layer of thin plastic can be placed on top of the previous layer. This is known as an additive process because each layer is added to the next. This is also how most home 3D printers operate.

How it Works

Many home 3D printers use 3 millimeter (mm) or 1.75 mm diameter ABS or PLA plastic spools of material that is routed through a gear drive mechanism that pushes the solid plastic line through a heated metal funnel. This reduces the size down to a soft stream of typically 0.4 mms wide molten plastic material. The heated mechanism is then controlled in the horizontal direction of both the X (width) and Y (length) direction to place the molten plastic on a flat bed. Some printers offer a heated bed with a glass top. The base gets lowered in the Z (height) direction by a very small amount, typically 0.1 – 0.4 mms and the next layer of molten plastic is laid down on top of the previous layer. This is called Fused Filament Fabrication.

An electronic module controls this operation by turning stepper motors very small amounts to control the X, Y and Z movement. The electronic module also has a USB connection that communicates with a computer that has the 3D sliced file. The file format is called a stereolithography file or .STL. This file contains all the encoded layer information. The file can be further reduced down to the X, Y and Z movements which is called a G-Code file. Some printers only accept the G-Code version and some accept the .STL and do the G-Code conversion themselves. Either way, most 3D designs start out as .STL files. The .STL file is then sliced by a software utility and then the resulting file is sent to the printer's electronics so the design can be printed in plastic.

3D Printing options

There are many different 3D printers to choose from. Some are in kit form and sold by individuals trying to create a business the same way Apple Computer and all the early computer makers started out selling computer kits, but that is quickly changing in the 3D market. Some of the big professional 3D printer companies have bought up smaller 3D printer companies or have designed their own home 3D printers so the competition is getting tougher. Here are some of the more popular 3D printers:

RepRap

This was the first home printer, which is an open source design that uses 3D printed parts in its design. The design is self-replicating as you need one 3D printer to make pieces for the next.

Makerbot

Makerbot improved on the RepRap design and made several versions of a home built kit you could buy. They eventually started making fully assembled units and finally were bought out by the larger 3D printing company Stratasys.

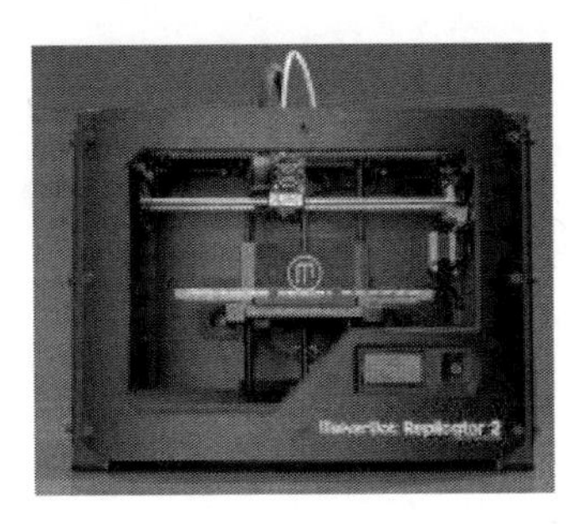

Cube

Cube was created by the professional 3D printer company 3D systems. There have been several variations to the Cube and they introduced the printer filament in a replaceable cartridge.

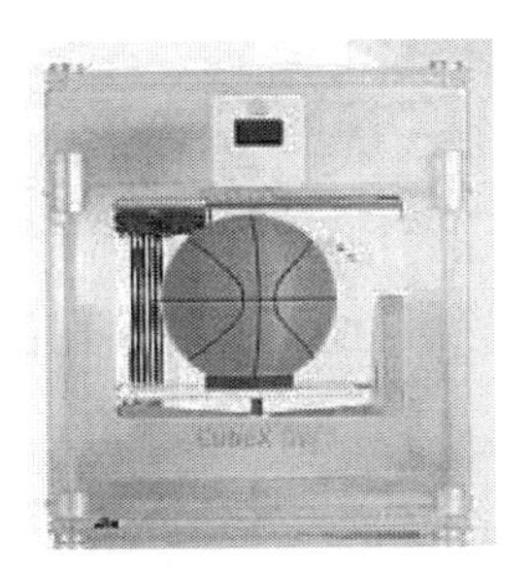

Solidoodle

Solidoodle was one of the first sub $500 3D printers and was started by former Makerbot employees. They offered a metal chassis instead of the typical wood or plastic frame. They have since released several more versions of the Solidoodle.

Printrbot

Printrbot started out as a kickstarter.com project that was extremely successful. They introduced a low cost printer kit. It quickly grew into a full time business and the low cost original design was improved but also increased in price significantly while they introduced new low cost designs to fill in the price gap. They continue to innovate and have moved from wood frames to metal.

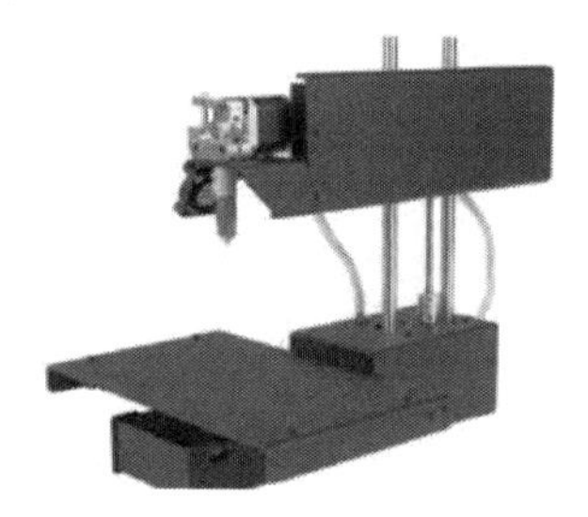

Da Vinci 1.0

The Da Vinci 1.0 was the printer I was looking for. I wanted a fully assembled unit and I wanted a large enough print area to build larger plastics. I also wanted something that was easy to use and didn't look like an incomplete structure. Most of all I wanted it to be under $500, if possible, and the Da Vinci met all these requirements and more. This book will assume the user has a Da Vinci 1.0 printer as it offers the best printer for the money and is the easiest to setup and use for the beginner just getting started with 3D printing. But almost everything in this book can be applied to any of the printers shown.

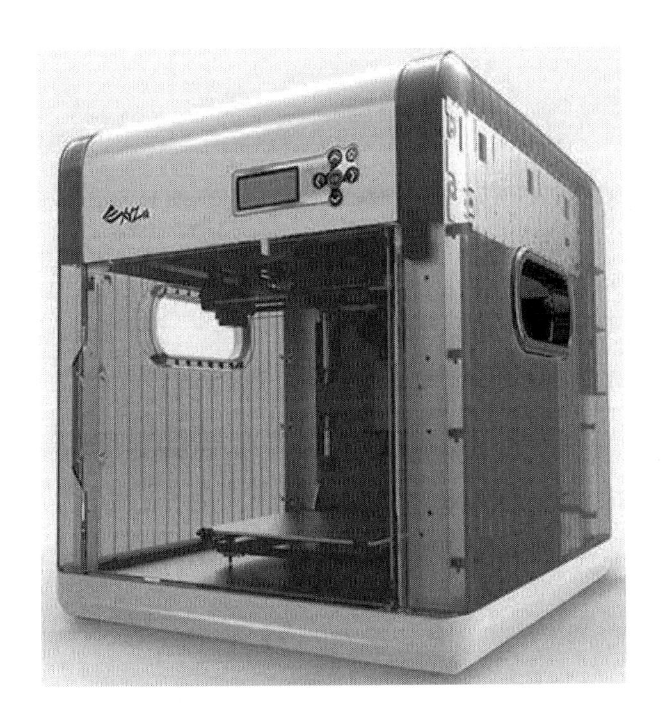

Da Vinci 1.0 Printer Specifications

Printing Technology:
FFF (FUSED FILAMENT FABRICATION)

Maximum build volume (WxHxD):
7.8W X 7.8H X 7.8D INCH (20 X 20 X 20 CM)

Printing Height:
0.1 MM (100 Microns) , 0.2 MM , 0.3 MM , 0.4 MM

Print Head:
SINGLE NOZZLE

Nozzle Diameter:
0.4 MM

Print Speed:
150 mm/s

Filament Diameter:
1.75 MM

Filament Material:

ABS (12 COLORS)

Filament Cartridge:

900g (N.W. 300g) / 240 Meters

Software

File Types:

.STL, G code, XYZ Format

OS Supports:

Windows XP (.Net 4.0 required), Windows 7, Windows 8.x

Mac OSX 10.8 64-bit +

Hardware

Hardware Requirements (FOR PC/MAC)

PC X86 32/64-BIT COMPATIBLE PCS WITH 2GB+ DRAM

APPLE MAC X86 64-BIT COMPATIBLE MACS WITH 2GB+ DRAM

Display

Panel Type:

4 X 16 LCD

Language:

ENGLISH, JAPANESE

Connectivity:

USB 2.0

Dimensions

Dimension with Spool(WxHxD)

WIDTH 18.4 INCH / 46.8 CM
HEIGHT 20 INCH / 51 CM
DEPTH 22 INCH / 55.8 CM

Net Weight (Cartridges included)

51.9 LB/23.5 KG

3D Printer Software

3D Printer software can bring up multiple confusing topics since there are different levels of software for 3D printers. There is the software that you use to create 3D objects on your computer. There is software used for slicing the object into layers. There is software to convert the layers into G-Code. Then there is software (sometimes called firmware) that is in the electronics that controls the 3D printer's X, Y and Z direction motors.

This was probably one of the more confusing topics for me when I started looking at 3D printers. It seemed like everyday there was somebody releasing a new version of software that they claimed was better than the others. But which category of software was it for, I often wondered? How would I use it? How was it better? And some of the released software combined the slicing with the G-Code creator. Some only worked with certain types of firmware in the 3D printer. So it became clear to me that unless you had some experience with 3D printing, this was just a confusing mess.

Replicator was the first G-Code creation software I found and it was used with the RepRap and Makerbot and was probably the most popular initially. It would load a .STL file and slice the file into G-Code that was directly sent to the electronics in the printer. But some people didn't like the slicer so they made a new one. And some added extra features. The software was open source so anybody could get the original files and make their own version. This is great for someone who wants to fool with all that but if you want to just use a 3D printer, all they were adding initially was more confusion. And I also noticed that the software to create the 3D object in the first place was rarely mentioned. There were references to Google Sketch-up but that required a plug-in. There were references to professional software but those were expensive. This was a whole new topic of software I needed to understand.

Since 3D printing has been around in the professional world, there are already many options for creating a 3D object design. These are called Computer Aided Design (CAD) software packages and most cost a lot of money. So I began a search to understand the CAD options and came across Tinkercad. Tinkercad is free software that you run online (through a web browser like Google Chrome) and you can build 3D objects in a way that is similar to building Lego blocks. I found the software easy to use but also very useful for creating 3D objects. The company was bought by Autodesk and is now part of a complete software suite, yet you can still use it for free. I'll explain Tinkercad more in a future chapter.

Da Vinci 1.0 Software

One of the selling features of the Da Vinci 1.0 printer for me was the software included to print a 3D object. The firmware that runs the electronics inside the machine was installed and can be easily upgraded from a laptop. The XYZ Printing software, XYZware, does a lot of what the replicator software does in that it loads the .STL file and then when you hit print, it will slice the object and send the G-Code to the printer. Best of all, it doesn't cost extra as it was included for free. In fact you can download it ahead of time from the XYZ Printing website (http://us.xyzprinting.com) and use it before you buy the printer, though you need the printer hooked up to do any printing.

This simplified it all for me. I didn't have to go searching around to determine which 3D printer to buy and which software to use with it. I could get most of it in one package and then when I wanted to create my own designs I could use Tinkercad.

For this book we will use these three pieces of software to create 3D printed designs.

Tinkercad + XYZware + Firmware in Da Vinci 1.0 printer

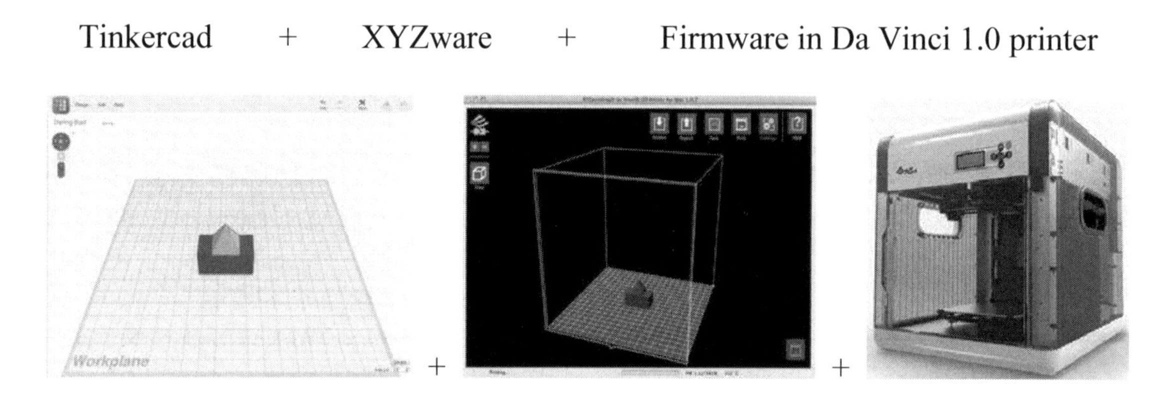

Sample Objects

When you first get started with 3D printing it helps to have a sample or two of objects to print. The Da Vinci 1.0 printer handles this as well. Stored within the Da Vinci 1.0 printer firmware are three sample files you can print. I'll walk you through this in the next chapter but know that you don't need Tinkercad or the XYZ Printing software to get started. You can start a print right from the LCD screen of the Da Vinci 1.0 printer.

Beyond those three samples, you can also download more samples from the XYZprinting company website. There are many samples there to choose from. You just download the .STL file and load it into the XYZ Printing software and then you can click on print for it to be sliced and sent to the Da Vinci 1.0 printer.

There are also numerous websites where people share and sometimes sell their 3D designs. The most popular is thingiverse.com. It has thousands of different objects to choose from. Most files will work on the Da Vinci 1.0 printer but you will need the .STL file and also make sure it was made in mms not inches. That is not always apparent so you may have to send a message to the creator. The original Da Vinci 1.0 printer required mm encoding in the .STL file.

Tinkercad is another source of files. Many people share their Tinkercad files on that site. This allows you to load the file and modify it if you want and then export the .STL file to the XYZ Printing software so it can be sliced and sent to the Da Vinci.

Professional 3D Printers

There are companies that accept your .STL file and will print it on a large professional 3D printer. They can also offer different materials including wood and metal for some prints. This adds a nice option as you can work out the bugs of your design at home and then have it professionally printed. The cost of these prints is a lot more than you pay for the plastic in your own printer. You can save a lot money making test prints with a home printer even if you only catch one error. It is nice to know you can take a design all the way to a professional print with the same tools and files you are using at home. Tinkercad has several of these companies already linked in so you can send it straight from that software.

Chapter 2 – Da Vinci 1.0 Printer

The Da Vinci 1.0 printer is actually quite large compared to other 3D printers. It comes fully assembled in a large box. It's packaged really well to help prevent damage during shipping. I spent almost 30 minutes just unboxing and setting up my Da Vinci. I'll step you through some of the important steps here.

Figure 2.1: Da Vinci 1.0 Printer Arrived

Unboxing

The box is heavy and has multiple straps to remove but when you remove them you can actually lift the top of the box off without cutting any of the cardboard. This helps because I was concerned I might cut something important inside the box.

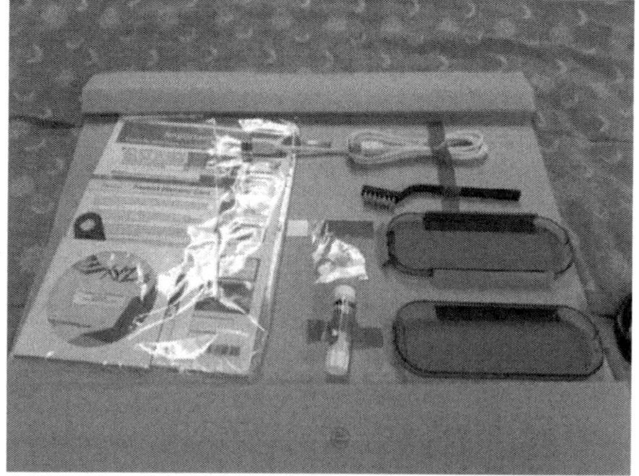

Figure 2.2: Plastic Cartridge and Accessories

Figure 2.3: The Top of the box lifted off

The box also contains a 300-kilogram cartridge of white plastic ABS filament (120 meters) and all the cables and tools you need to print. Also included is a setup manual and a CD with the software though I recommend downloading the latest from the http://us.xyzprinting.com website.

There are also two oval shaped plastic parts that are the handle inserts for the sides of the Da Vinci. A glue stick for helping to hold down the prints is included and a wire brush for cleaning the extruder (the heated metal funnel where the plastic comes out).

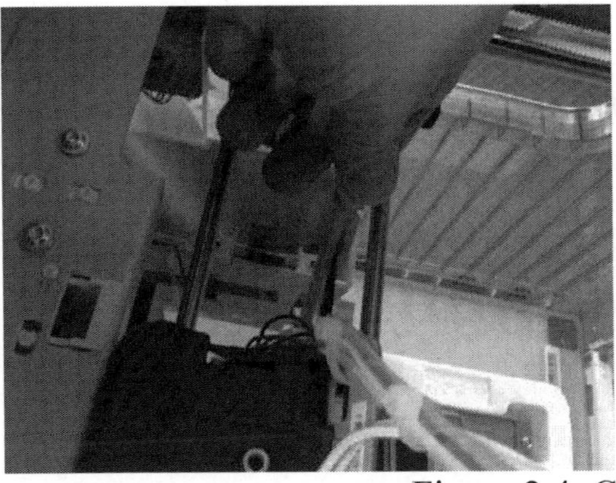

Figure 2.4: Clips to Remove

Inside there are a couple of clips to remove through the top of the unit and then there is a base clamp that is screwed down that needs to be removed. Funny thing is these brackets could have been 3D printed but they weren't. There were also a lot of cardboard inserts and Styrofoam that needed to be removed as well. Over all the unit was well protected from damage during shipping. There was also a lot of tape to remove. A lot of tape!

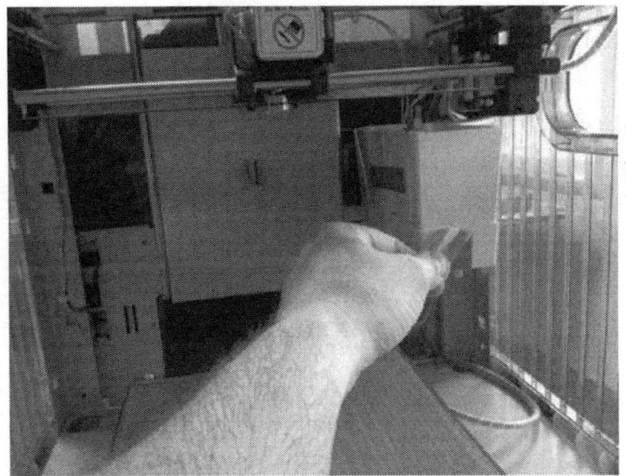

Figure 2.5: Packaging to Remove

When it was all unpacked I could see that the Da Vinci 1.0 printer was really well built and quite large. For the $499 price is was a great deal in my mind but the real test was the first print.

Installing Filament

The filament comes in a cartridge that needs to be installed.

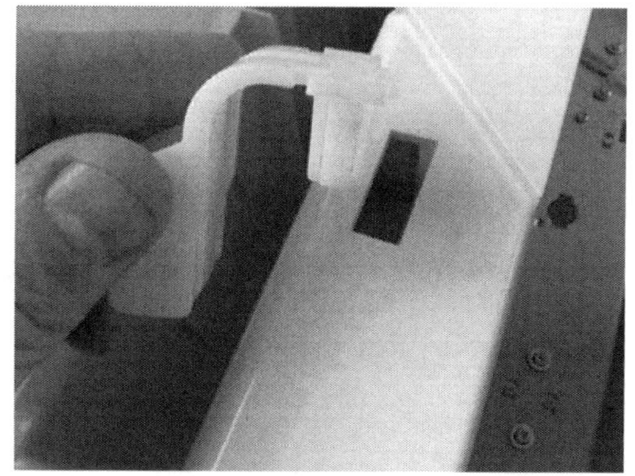

Figure 2.6: Preparing the Filament Cartridge

The filament cartridge has a small plug that needs to be removed and the printer has a lock-down bracket the also has to be removed. You open the top door of the printer to get to the lock-down bracket. There you will see the slot for the cartridge to be installed. It just slides into place and then is locked down by re-installing the lock-down bracket.

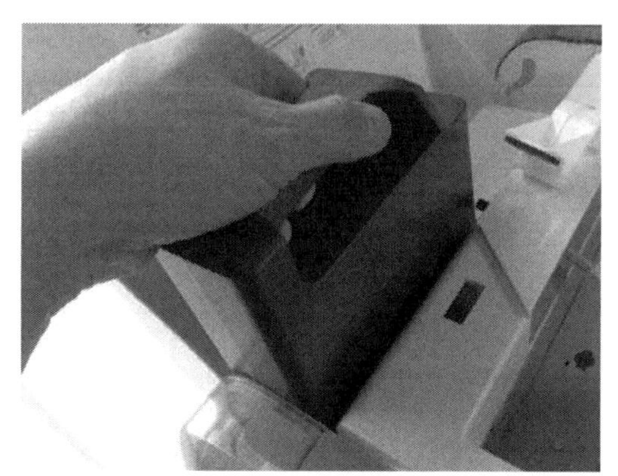
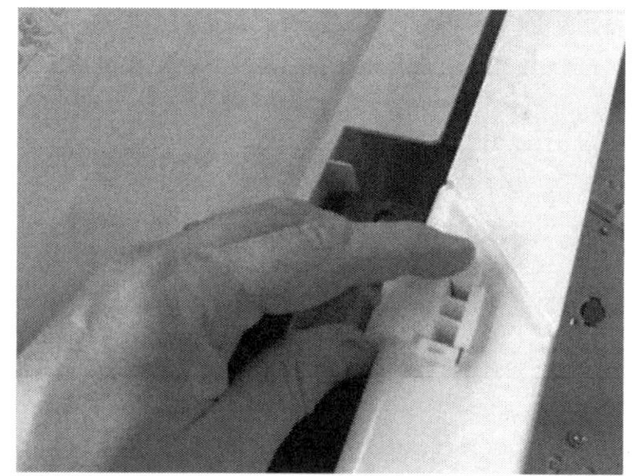

Figure 2.7: Installing the Filament Cartridge

Now the filament has to be fed through the tubing and into the extruder. This is easy to do but very important to get right. The extruder has a spring-loaded lever that needs to be pulled out so the plastic filament can slip between the gears that will pull it in during prints. This can be a bit difficult if you have short arms. I have long arms and it was still hard to reach in and feed the plastic in with one hand while pulling the lever out with the other. Feed the plastic in until it hits a solid bottom. That solid bottom should be the extruder metal funnel.

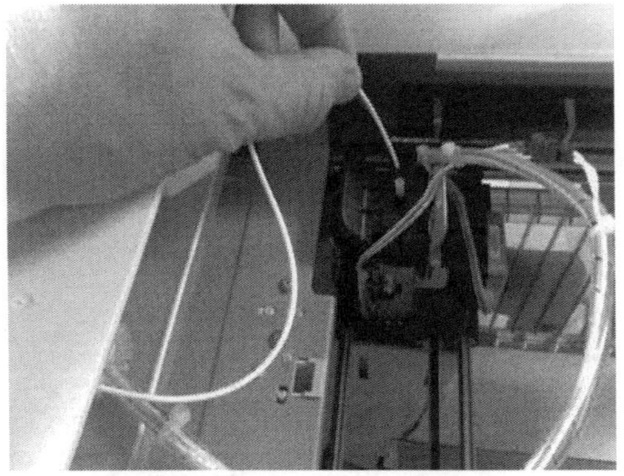

Figure 2.8: Routing the Filament

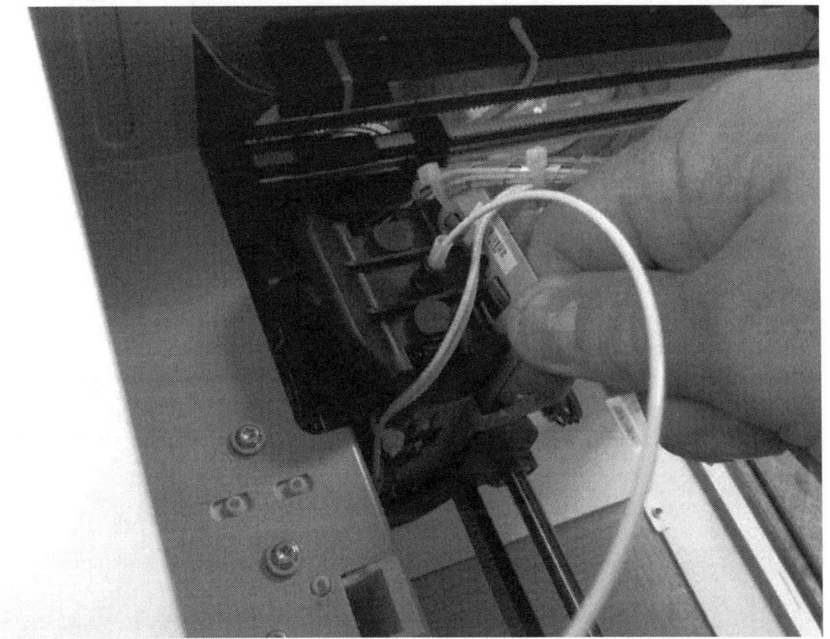

Figure 2.9: Releasing the Spring-Loaded Filament Router Gears

Now the printer has to be powered up so the motors can drive the filament into a heated extruder funnel. Plug in the included power cord and then press the switch to turn it on. The Da Vinci should start making some noise and you should see the LCD display light up with a menu of choices to select from.

Figure 2.10: Connecting Power and Setting Power On

Figure 2.11: Da Vinci Powered Up

Just placing the filament into the tube is not the final step; it needs to be routed into the heated extruder nozzle. The software handles this so the first step, with the arrow pointing to the UTILITIES option on the LCD, is to click the OK button. You will see the next menu appear.

Figure 2.12: Load Filament Menu

The extruder will begin to heat up to about 210 degrees C and then automatically the motor in the extruder will start pulling in the plastic filament and begin pushing it through the heated funnel of the extruder.

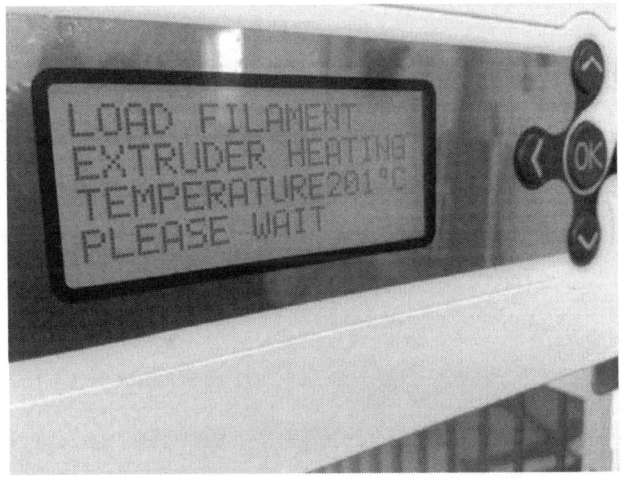

Figure 2.13: Extruder Heating Up

The plastic bucket below the extruder will collect the plastic as it squirts out. This will run for a short time and then automatically stop. I positioned the extruder so the plastic would collect on top and I could show how it looks.

Figure 2.14: Extruded Plastic

The plastic cartridge that comes with the Da Vinci 1.0 printer has white ABS plastic inside. It appears that my machine was tested at the factory with red because, as you can see, there is red in the first part of the extruded plastic and then eventually becomes white. The tornado shaped object in Figure 2.13 shows the extruded plastic. You could call this your first official 3D print!

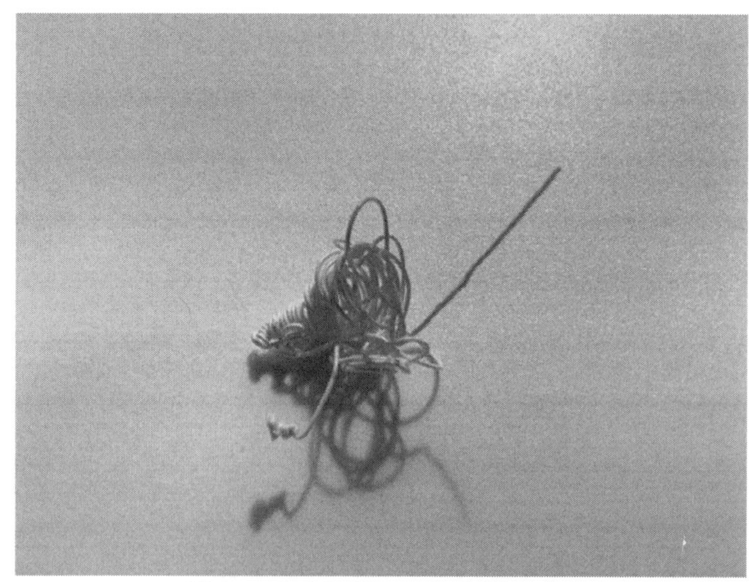
Figure 2.15: Extruded Plastic

This procedure will be used when you change cartridges so recognize that you have the option to run it a second time if you need to get more of the old color out of the system. After the extrusion stops it asks if you are ok to continue (Return) or want to extrude some more to get the previous color out. Click on OK to move on.

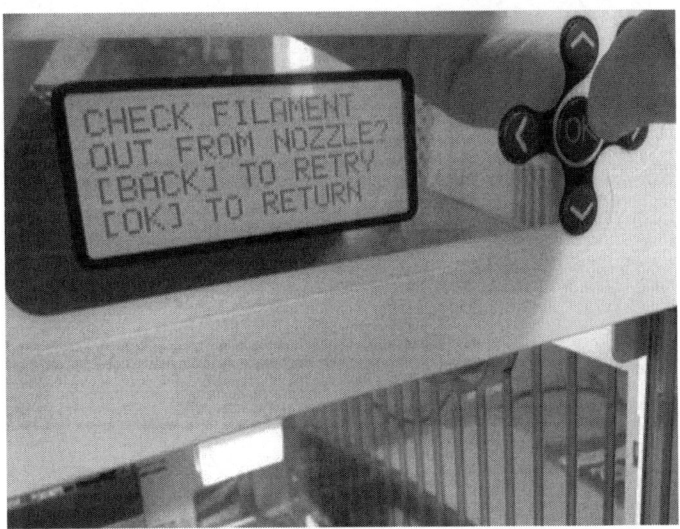

Figure 2.16: Final Step to Load Filament

The printer is now setup and ready to make your first 3D print.

Chapter 3 – First Print

The Da Vinci 1.0 printer is unique in that you don't have to connect a computer to create your first print. The printer has three sample designs stored in its memory so you can test it right after setup. I'll step you through your first print here.

Figure 3.1: Keychain Sample Print

The first sample we will print is the XYZ keychain logo. It has a hole in a tab for the metal key ring to go through that actually holds the keys. Its the smallest of the three samples and a great place to start.

Printing the XYZ Keychain

The Da Vinci 1.0 printer includes a small glue stick. This is to create a sticky surface for the first layer of plastic to grip onto. Getting the first layer to stick is critical to getting good prints out of the printer. This is why the height adjustment of the heated glass base is so critical to get right, and why I cover it in a later section of this book. Your new machine should have the base already adjusted at the factory so don't touch the thumbscrews at the bottom of the base cause that will mess it up.

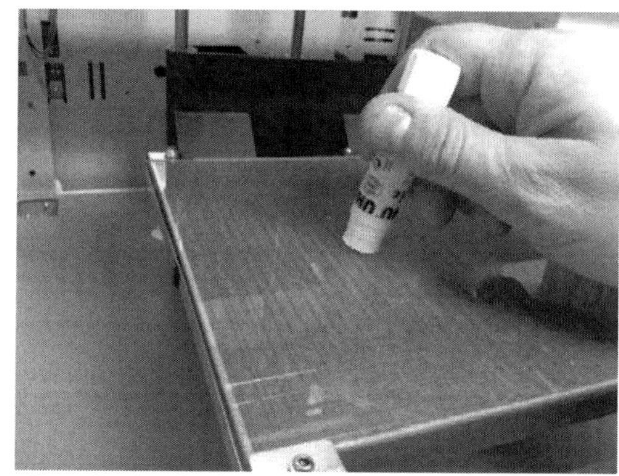

Figure 3.2: Gently Rub the Glue Stick onto the Platform Base

Rub the glue stick onto the center of the base. You can cover a large area without any problems. The glue stick is recommended because it cleans off easily with water. Some people recommend hair spray but it hasn't worked very well for me. I think the brand is important and I'm not sure which one to recommend. I have found that Avery Glue Stick brand works best as it's washable and also holds the print better than the glue stick included with the Da Vinci 1.0. If the platform is adjusted right and the build is small, I've found I don't need a glue stick and the prints come out very clean and so does my glass platform.

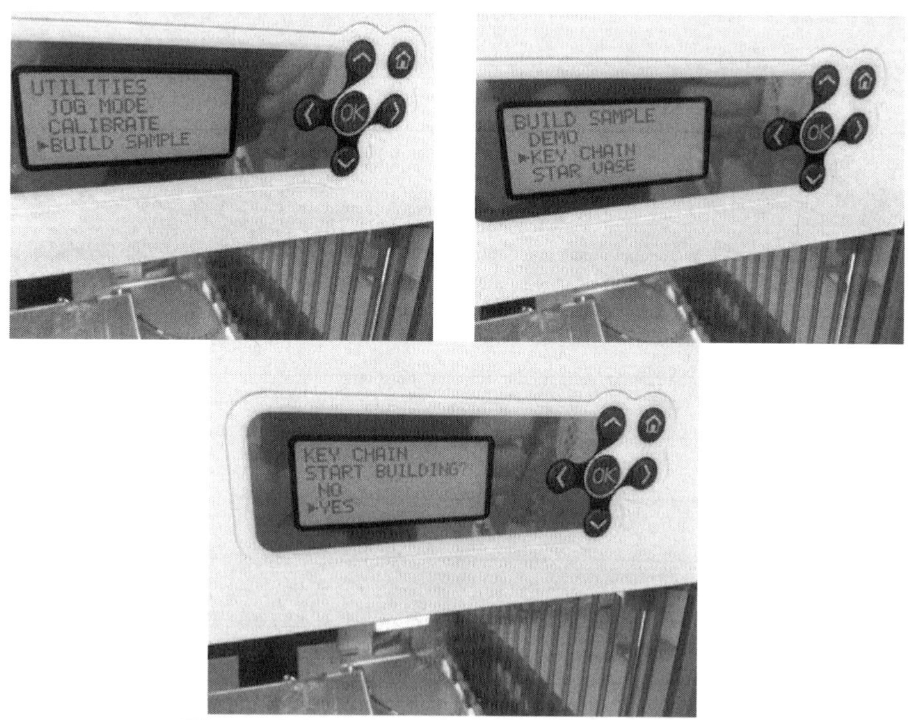

Figure 3.3: Select the Key Chain Sample

The sample is selected from the Utility menu. The Utility menu is longer than the screen so if you click on the down arrow you will eventually see a BUILD SAMPLE option. Click OK and then select the KEY CHAIN. The menu will then give you one more chance to change your mind. Click YES to start the print.

Figure 3.4: Platform Rising

The Build platform will begin to rise up. Then the extruder and the heated glass top platform will begin to heat up. This will take a few minutes. When the temperatures are achieved (about 210C for the extruder and about 90C for the platform) the build will begin. The extruder will move across and back as it places the plastic in the proper spots. As each layer is completed, the platform will lower a little bit. This can be anywhere from 0.1 mm to 0.4 mm depending on the settings in the file.

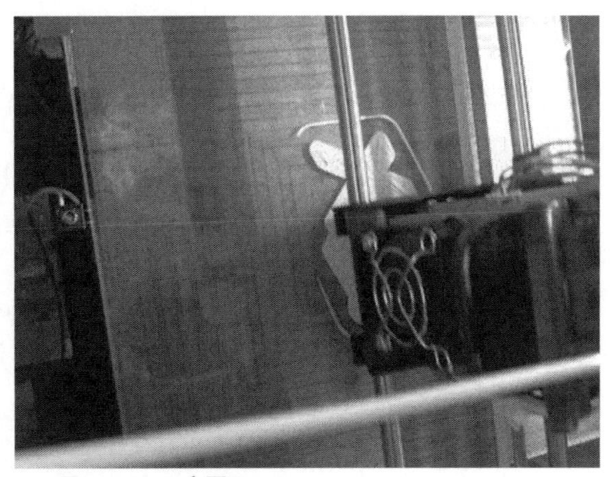

Figure 3.5: View from Front and Top

The LCD screen will show the progress as a percentage. It will also show the total time and how much time has passed. This is not always perfectly accurate but it's close.

Eventually the print will finish but the platform will have to cool down. The LCD will display when the print is ready to remove. The platform will also lower automatically so you can get to the object easily.

A plastic putty knife is included to help you scoop up the print without scratching the glass base. As you can see, the demo print actually printed facing away from the front of the machine.

Figure 3.6: Remove Print from Base

Figure 3.7: Completed Sample Print of XYZ Keychain Logo

Chapter 4 – Thingiverse

The next project to try is to print somebody else's 3D print design. There are many different websites that list 3D designs that people share open source. This allows you to print the design and also modify it if you have the software to make those modifications. I'll discuss one version of that software in a future chapter, but for now we'll use a file I found on Thingiverse.com.

Thingiverse was one of the first, if not the first, location for people to share their 3D prints. XYZ Printing also has a website with free 3D print files you can build, but I find their lack of a search engine and the crude pictures of some of the 3D prints to be a bit difficult to work with. Thingiverse also has been around a lot longer so it has many more choices.

The first print I went looking for was a charging stand for my iPhone. I did a search and found many different designs but decided to build the one from user named *schrotti*. The design was for an iPhone 5 or 5S and I have an iPhone 5. The design has a channel designed in for the charging cable and it appeared to hold the phone in a way that was easy to install or remove the phone from the charger.

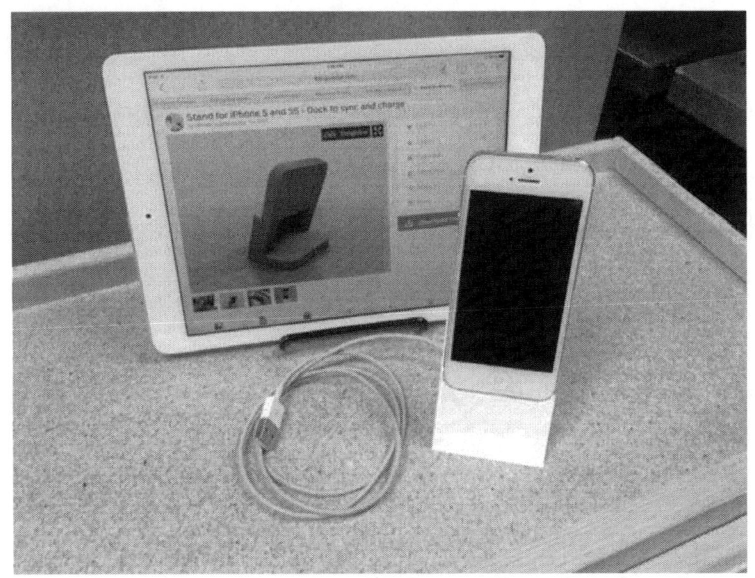

Figure 4.1: Completed iPhone Charging Stand

Most of the files on Thingiverse are .STL format or STereo Lithography format. This is one of the original 3D printer file formats for professional 3D printers. An .STL file contains all the design detail in a layer-by-layer format so it is easy to slice into layers for 3D printing. The XYZware software has built in slicer software, which takes the layers in the .STL file and converts it to the G -Code the Da Vinci printer needs to control the print head in the X, Y and Z directions. G-Code is another common format and is used in various forms for cutting 2D images with a CNC cutter. A form of it is even used to cut circuit boards. You can read a G-Code file with a simple text editor if you know all the custom codes.

So, the first thing we need to do is download the .STL file of the iPhone stand from Thingiverse.com. The site will have a menu item that says "Thing Files". Click on that menu item and the list of .STL files will appear ready to download. In this example there is only one file to download, as seen in Figure 4.2.

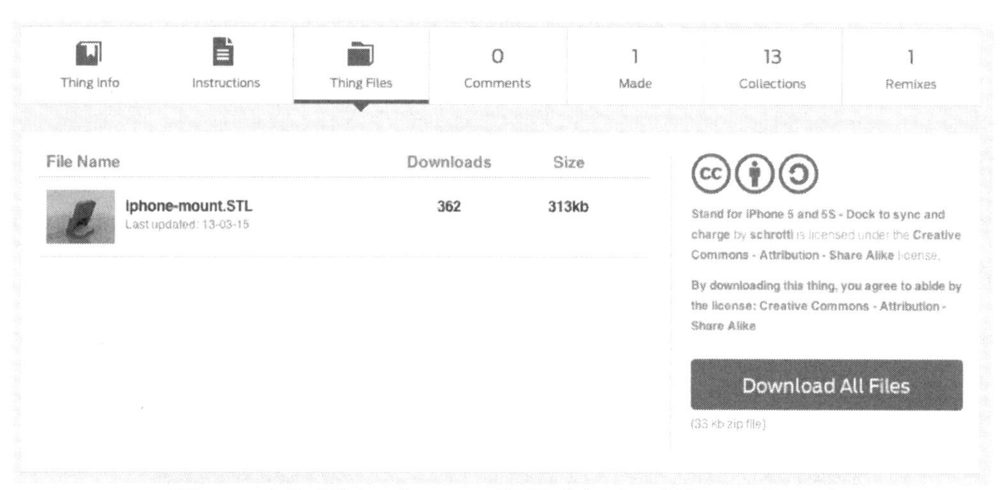

Figure 4.2: Thingiverse .STL file for iPhone Stand

Just click on the picture or click on the "Download All Files" to get the file for printing. Store the file in a location where you can find it because we'll load it into the XYZware software a little later.

Note: One thing to be forewarned about is the early Da Vinci 1.0 printer worked with files dimensioned in mms. Sometimes the Thingiverse file will be in inches so when it is loaded in it can be incredibly small since the XYZware will think the dimensions are mms. I've only run into this on rare occasions but be careful because it's not always specified on the Thingiverse page.

Once the .STL file is downloaded, then we can launch the XYZware used to send prints to the Da Vinci 1.0 printer. We didn't need this software for the sample prints but we will need it for this project.

XYZware Software

The XYZware comes with your Da Vinci 1.0 printer on a CD, but it is recommended to just download the latest version from the Da Vinci website (us.xyzprinting.com). You can click on the Download menu item but you will be asked for a login and password to get to the download files. I suggest you create a login, as it's free, and any contact with them for support may require it.

After you login you will see a screen with software downloads for both Windows and Mac. There is also a firmware download option but that is for updating the internal code that runs the printer, not the software you run on your computer.

XYZware software is used to slice .STL files into layers and then convert that information into a G-Code file and send that to your Da Vinci 1.0 printer. XYZware does not allow you to create designs or even modify them other than to resize them. To create custom designs you need a CAD or Computer Aided Design software package. This can get confusing for a beginner, but I'll show you an easy to use CAD software called Tinkercad in the next chapter.

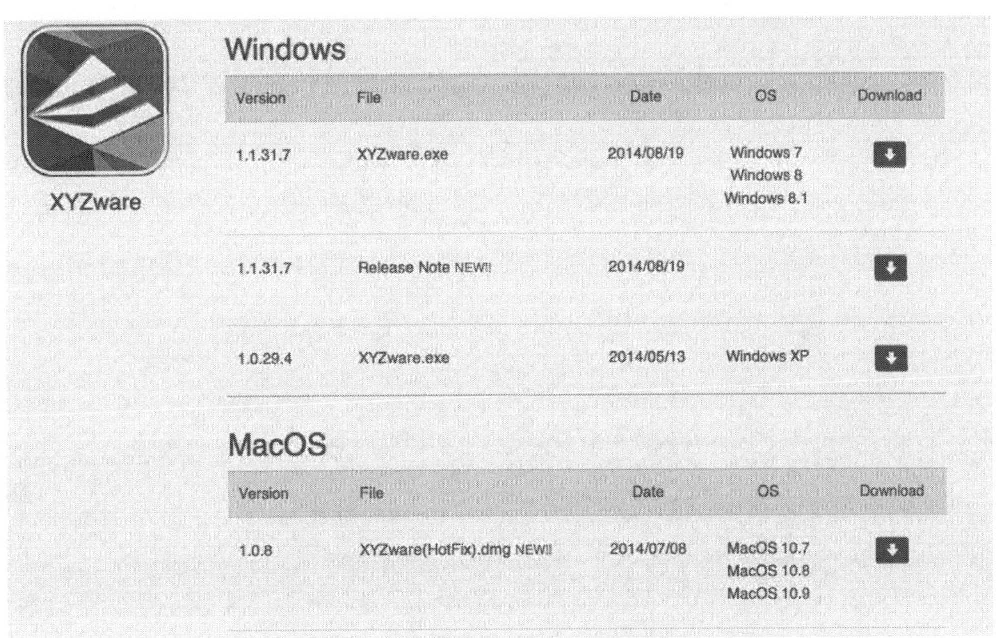

XYZware

Windows

Version	File	Date	OS	Download
1.1.31.7	XYZware.exe	2014/08/19	Windows 7 Windows 8 Windows 8.1	
1.1.31.7	Release Note NEW!!	2014/08/19		
1.0.29.4	XYZware.exe	2014/05/13	Windows XP	

MacOS

Version	File	Date	OS	Download
1.0.8	XYZware(HotFix).dmg NEW!!	2014/07/08	MacOS 10.7 MacOS 10.8 MacOS 10.9	

Figure 4.3: XYZware Download Options

Download the XYZware version for your computer and install it. I find the Windows versions have the latest features while the Mac is close behind. I've actually run the software on an older XP based netbook and it ran fine.

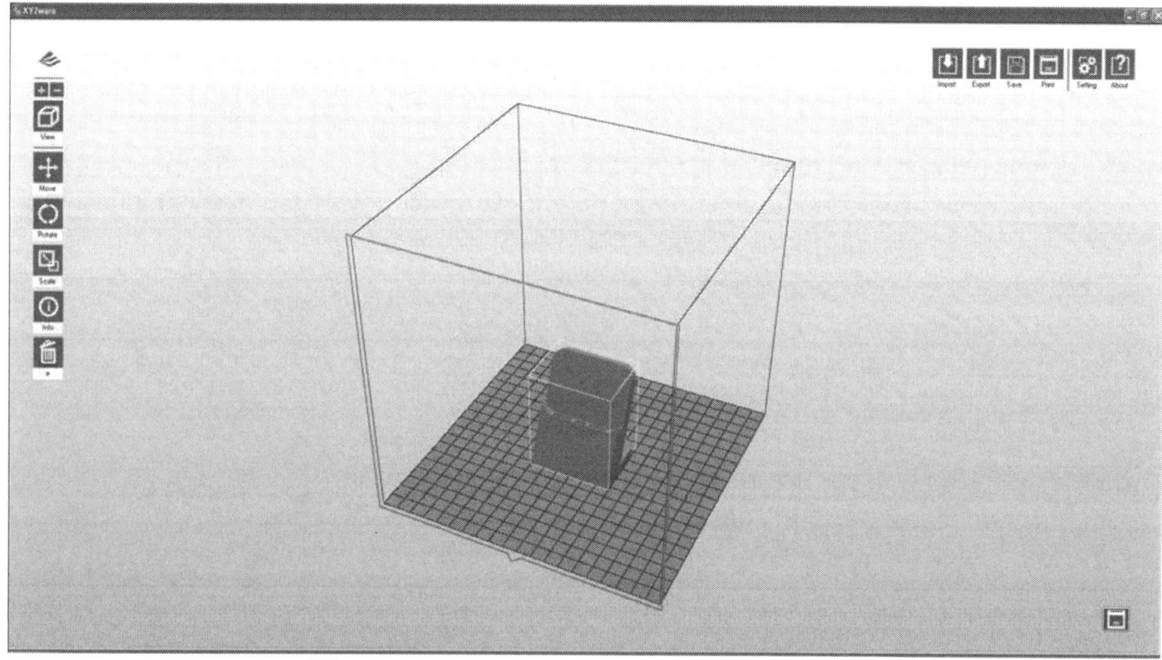

Figure 4.4: XYZware OSX Screen Capture

For this example I'll step through the Mac OSX version of the XYZware options but these same options appear in the other operating systems, though they may look a little different.

There are two menus of icons that make up the XYZware control: the menu on the left side of the screen and the menu at the top. The menu at the top is where you can load and export a design.

TOP MENU

IMPORT

The import option is how you load in a .STL file. A file finder window will open and you navigate to the file you downloaded from thingiverse.com, and it will drop into the XYZware platform. You can import more than one but you will have to move them separately.

EXPORT

The export option will let you select the slicing options such as height of the layers, fill ratio, and if you want the design built on a layer of plastic (called a raft) or not. You can also enable supports so the design will print plastic under any overhanging parts of the design. When the design is set the way you want you click on the export button and the G-Code file is created.

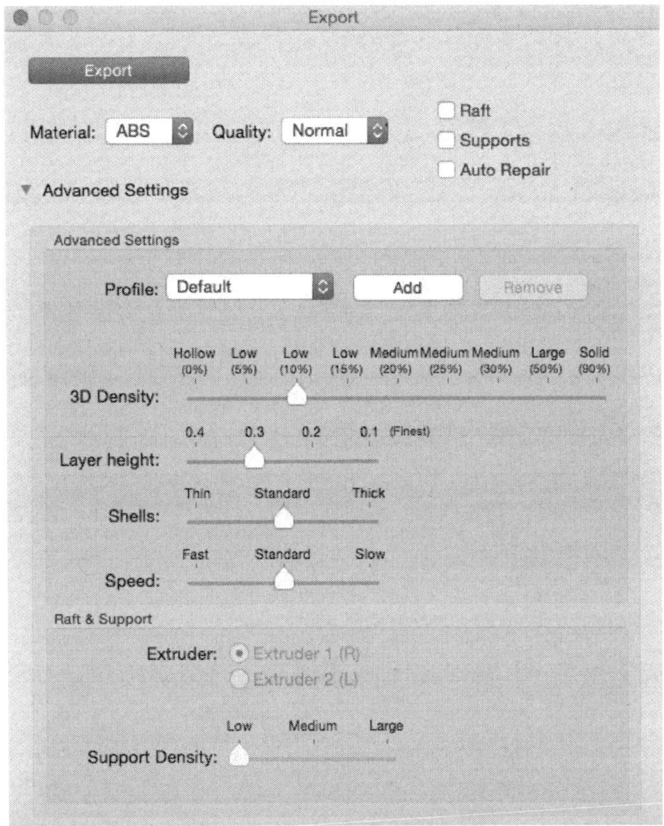

Figure 4.5: Export Screen

The settings are the most important part of sending your design to the 3D printer. The plastic selection shows ABS or PLA. The early Da Vinci 1.0 printers could only print ABS but PLA cartridges are scheduled for release by February 2015. Select the Excellent (0.2mm layer), Normal (0.3mm layer) or Good (0.4mm layer) setting in the Quality drop down menu. I find Excellent is a great place to start any print. You can also click on the Advanced Settings and control the settings directly.

There are three other options for a print before we cover the Advanced Options.

Raft - This puts a layer of plastic down across the whole width of your print. This helps the design stick to the heated glass bed. It works great but is all a pain to remove from the bottom. Use this only if the bottom of the print can be sanded or filed away.

Supports – This is a very handy option that will build a tower of plastic under any portion of the design that is floating or has lots of empty space underneath it. Supports are printed to easily breakaway and can be made in three different levels of density but the lowest is the easiest to breakaway.

Repair – This feature tries to repair your .STL file. Sometimes the XYZware has trouble reading all the .STL layers. The result will be an incomplete imported design. This feature, if checked, tries to correct that with an internal algorithm. Sometimes it works and sometimes it doesn't.
Another option is the Microsoft file repair facility at netfabb.com. It's free and will export a G-Code file that you can import into XYZware.

I've had some success with either of these options but some files just don't import.

ADVANCED SETTINGS

3D Density – This determines how much plastic each layer will contain. The layers are built in a criss-cross pattern and the density setting in the build will set the spacing. The picture below shows a 50% fill.

Figure 4.6: 50% Fill on 3D Printed Object

Layer – The layer is the thickness of each plastic layer put down by the extruder. The smaller the value, the more accurate the design print will be. But it will also take a lot more time if the layers are smaller. 0.2mm is the layer height used in the excellent setting of the XYZware and this prints really well while saving some time and plastic. I recommend starting at this level.

Shells – Shells determine how many runs the outer layers of a feature will get to make it stronger or thinner. You have three settings, which will produce one (thin), two (standard) or three (thick) passes to form the shell of the outer layer or around any holes in the design.

Speed – Sets the head speed for printing. This can affect the quality, so use high speed only when necessary.

Support Density – Just like the density of the design you can also set the density of the support if that is enabled. You want the opposite for supports as you do the design. Supports need to break away easy from the finished print so make these as low a density as possible. Starting at the low setting is typically all you need.

When you have everything the way you want it, click on the Export button to create the G-Code file and then it will automatically ask you for a file name to save it.

SAVE

Save is just a quick way to save any changes you make to the design.

PRINT

Print is very similar to the Export option. It will take a .STL file and slice it per all the same settings as Export, but then when you are done you click on the Print button and the file will get sent to the Da Vinci printer to be built.
The biggest difference between Export and Print is that Print needs the printer connected to your computer or it won't launch the settings menu. Export, on the other hand, can work with the printer disconnected. I often use Eport to create the G-Code file which will have a .3w suffix and then save it to a flash drive. I can then take the flash drive to a computer connected to the Da Vinci printer and send the file to the printer for the build.

SETTINGS

Settings lets you select automatic placement of a loaded .STL file and also the language used in the software. You can also select the Da Vinci printer you are connected to. There is a new 2.0 version that will be released before this book is finished that will have two print heads for dual color prints.

HELP

Help will give you a link to the online help at the XYZprinting.com website.

SIDE MENU

The side menu controls a print once it is loaded through the import menu. At the top are + and – icons for zooming in and out of the view. Using the scroll wheel on your mouse will do the same thing.

View is the first option to select. It allows you to choose the angle to look at the print.

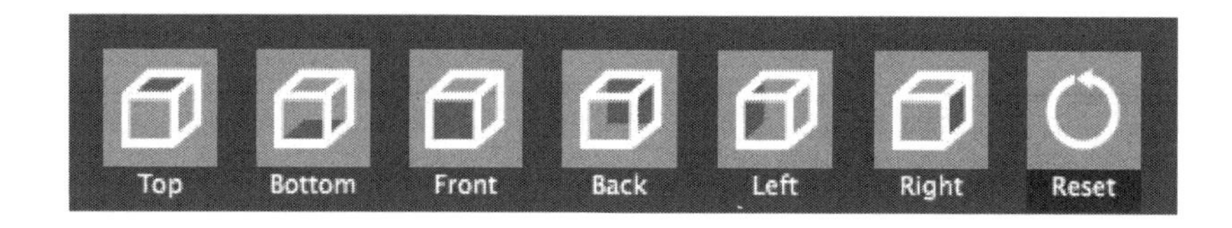

Move is next and allows you to move the object in the X, Y or Z direction. You can also Reset it to the original position and make sure its flat on the build surface by clicking on Land.

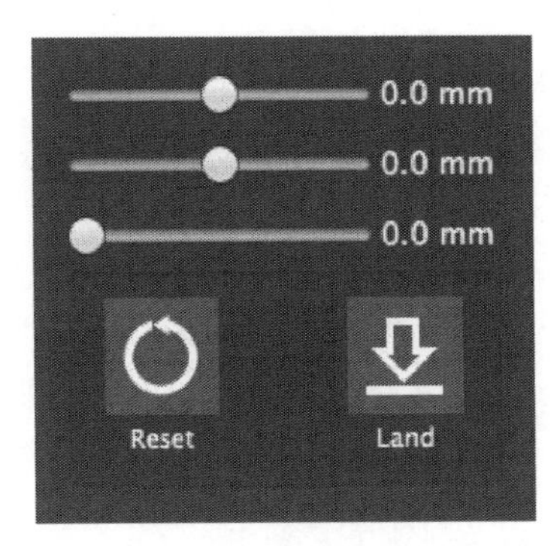

Rotate allows you to change the direction the print will face.

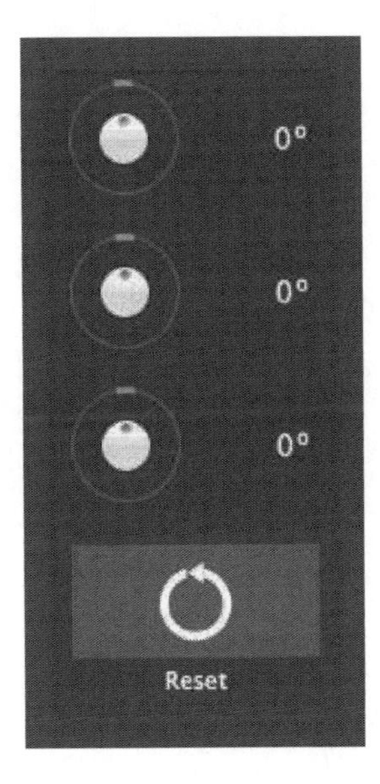

Scale allows you to resize an object as a percentage of the original. This can be very handy to make a smaller version of something you found on thingiverse.com that is bigger than you wanted.

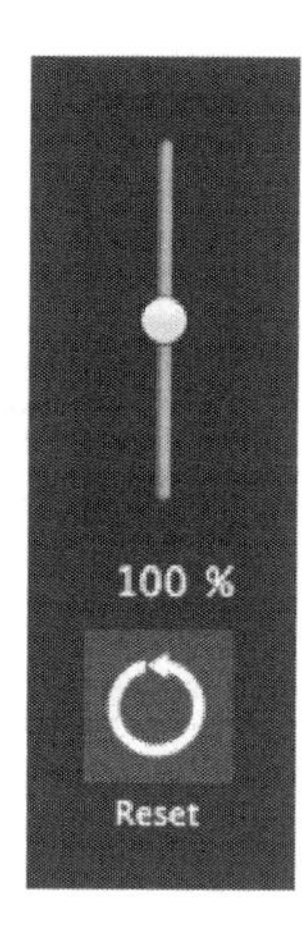

Those are the options XYZware offers for building your 3D print. You can download XYZware and use every option, except for Print, without a Da Vinci printer connected to your computer. This allows you to try out the software even before you buy a unit.

Some people find XYZware too simple and would like more control over the settings. Many other 3D software interfaces do offer more settings. Some are open source and constantly improving but they aren't connected directly to any 3D printer company so they can require a more complicated setup. The XYZware and Da Vinci were designed to work together and make it easy to 3D print something and they do that well.

Now you can send your Thingiverse file to the printer. If you downloaded an .STL file like the iPhone charging stand, import it into the XYZware and adjust its position if you think it needs to be moved or rotated. Odds are it's fine in the default position. Make sure you're connected to your Da Vinci 1.0 and then click on the Print menu. Select the Excellent setting and click on the Print button. You will see the design begin to be sliced. When that is done it will be ready to send to the printer. The stats for the print will show you how much time it will take and also how much plastic it will use of the existing plastic on the spool. The window shown below is actually from a Windows XP computer.

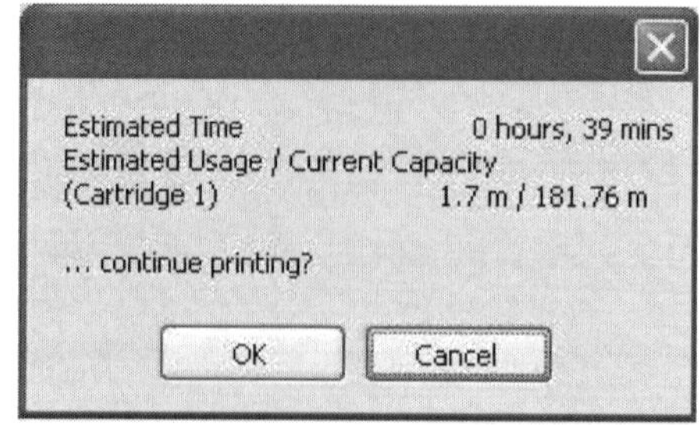

Figure 4.7: Windows XP Print Stats

Click on the OK button and you will see the download screen appear as it sends the file to the printer. In a few minutes, after the file is sent to the printer and the print bed and extruder are heated up, the Da Vinci will begin printing your design.

Figure 4.8: File Being Sent to the Da Vinci 1.0 Printer

When the design is done you can remove it from the printer and then use it just like I did with the iPhone charging stand shown in Figure 4.9.

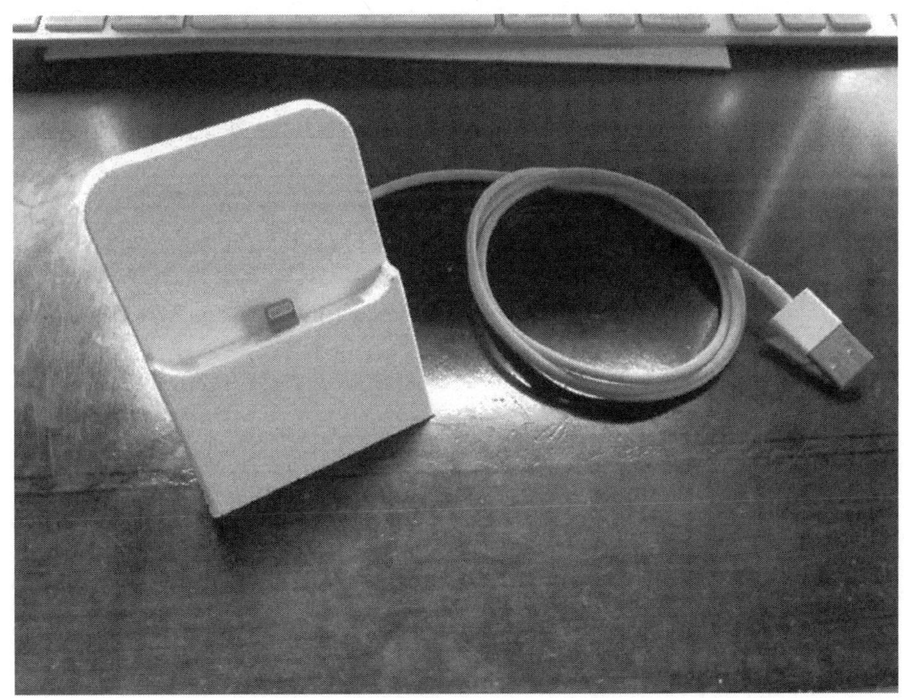
Figure 4.9: iPhone Charging Stand with Cable Installed

Chapter 5 – Tinkercad

This book will briefly explore Tinkercad for creating 3D designs, but recognize that Tinkercad will become one of your most useful tools as you get deeper into using a 3D printer. It is a simplified Computer Aided Design (CAD) tool that produces results even a professional could use. What Tinkercad gives you a simple way to build an object with blocks and shapes. You can adjust the size of the blocks and shapes while you build. You stack, place and shape the various pieces to create the object you want and then when you're done; you highlight all the different elements at one time and hit the group button to combine all the shapes into one solid design. You can then export a .STL file for the object that can be sent to any 3D printer. It's really like building an object with Lego blocks but in software, but unlike Lego blocks you can adjust the size and shape easily.

This book is not intended to be a complete course on using Tinkercad. That is left up to Tinkercad itself. Tinkercad is a free to use, online software tool that allows you to build any design and you can access it from any computer. Included within Tinkercad are tutorials that step you through various designs, which help to teach you how to use Tinkercad. Tinkercad also has a public sharing feature. Once you create a design, you can open it to share with the Tinkercad community so others can use or modify your design to make their own 3D design. You can also borrow from others and modify the design to fit your needs.

In this book I'll step you through one of the tutorial designs so you can work along with the book and get comfortable designing with Tinkercad. Then you can take that design and print it on the Da Vinci printer.

To run Tinkercad, Firefox or Chrome browser is recommended. Internet Explorer and Safari don't have all the required built in features to run the Tinkercad software. I recommend Chrome, which is a free download. Install it and then go to Tinkercad.com.

Once at the Tinkercad screen, create a login and then look for the "Lessons" menu on the left hand side under the COLLECTIONS heading. Click on the lessons link and you will see a bunch of lessons you can build. I'll step through the Chess Pawn build here.

Chess Pawn Tutorial

The chess pawn tutorial is a great place to start. It uses many of the most common shapes and has you resize and reposition blocks to make a pawn piece for a chess game. When it's done and printed it will look like the picture below.

Figure 5-1: Chess Pawn Piece

Click on the Begin Lesson button and the screen below will appear. This is the blank canvas where you will build your design. A pop-up window will appear and guide you through the steps to build the pawn.

Figure 5-2: Blank Start Screen

Step 1 - Place a cylinder in the location highlighted. Each step will highlight the outline of the piece to place on the platform.

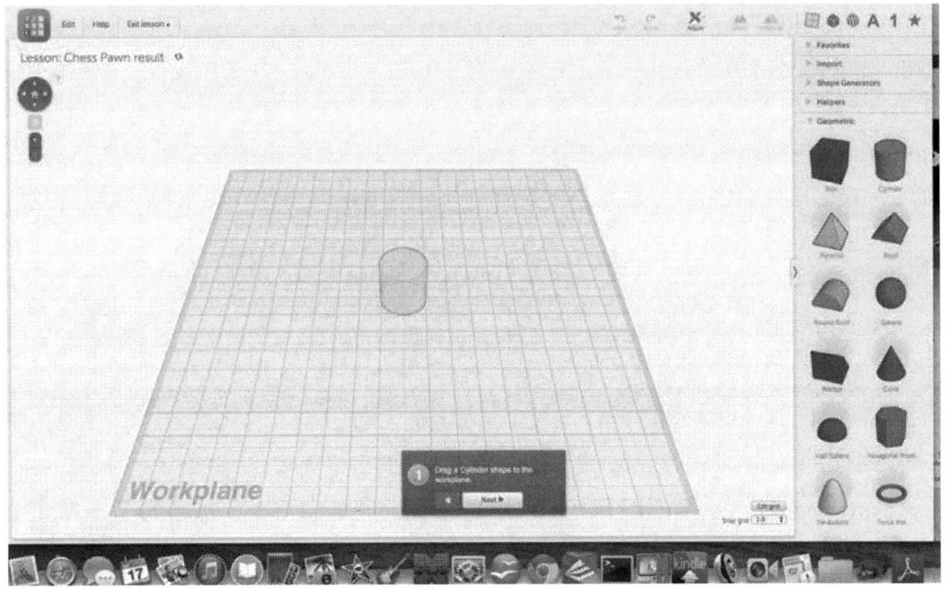

Figure 5-3: Step 1 of Pawn Build

Click on the orange cylinder from the right hand side and drag it to the outline on the screen.

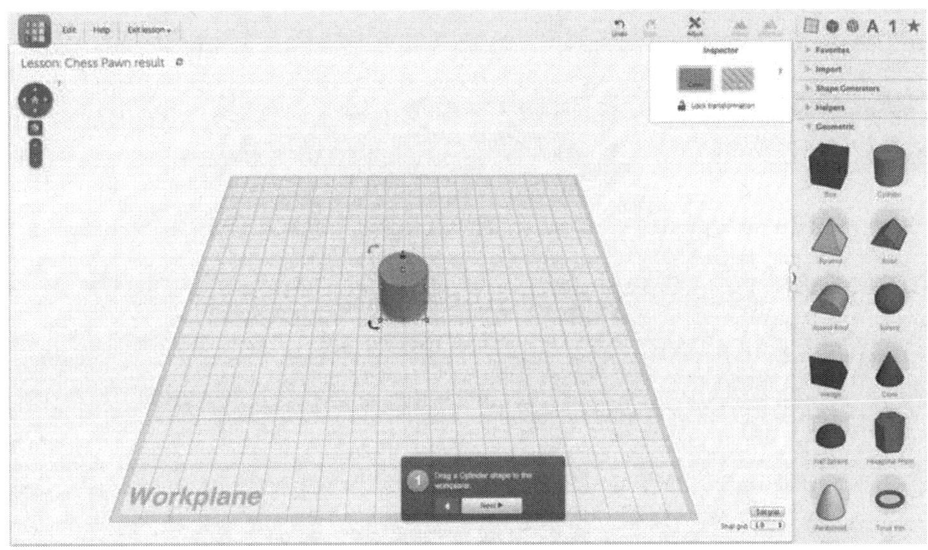

Figure 5-4: Step 1 Cylinder in Place

Step 2 - Reduce the height of the cylinder to 3 mms. Just click on the top square of the cylinder markers and while holding down on your mouse, drag it down until the measurement says 3.00 then let go.

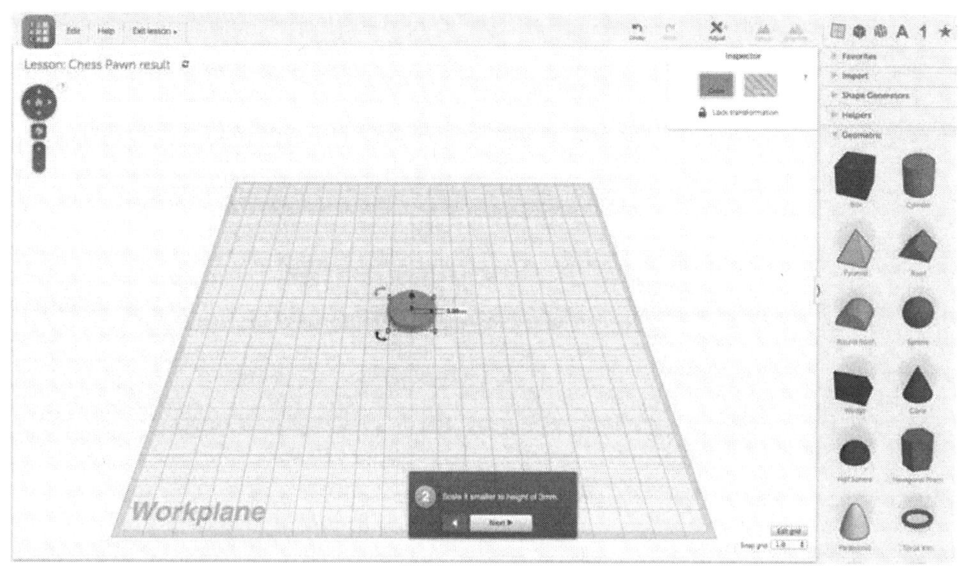

Figure 5-5: Step 2 Lower Cylinder to 3 mm

Step 3 - Move the build platform to the top of the disk you just created. You do that by dragging the Workplane, under the Helpers menu on the right, to the top of the disk. This will move the working platform to the top of the disk.

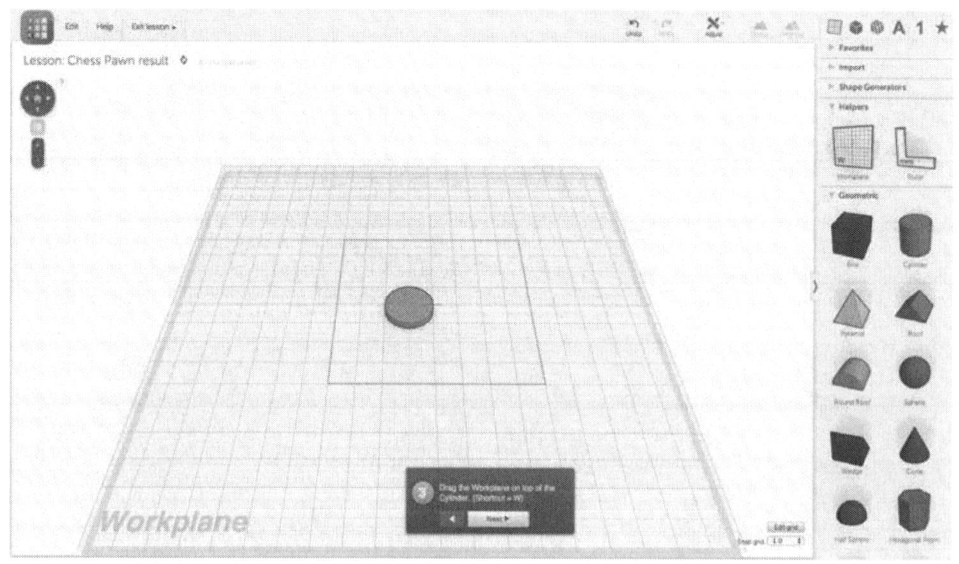

Figure 5-6: Move Workplace to top of Cylinder Disk

Step 4 - Select a Cone shape and place it on top of the disk.

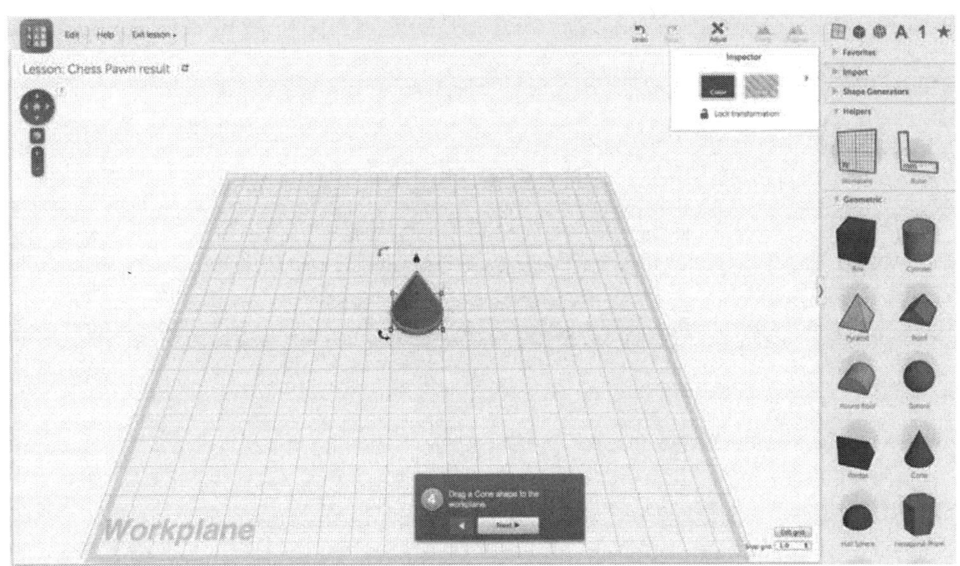

Figure 5-7: Add Cone on top of Disk

Step 5 - Reduce the height of the cone to 11 mm tall. You do this by clicking on the box element at the top of the cone and dragging it down until the height is 11 mm.

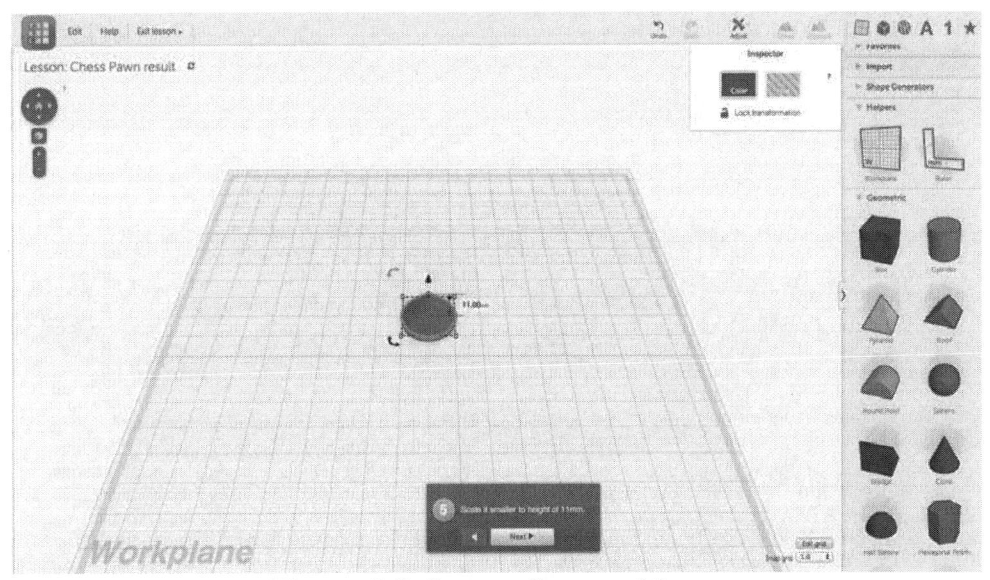

Figure 5-8: Lower Cone to 11 mm

Step 6 - Bring a sphere into the design and place it within the outline marker.

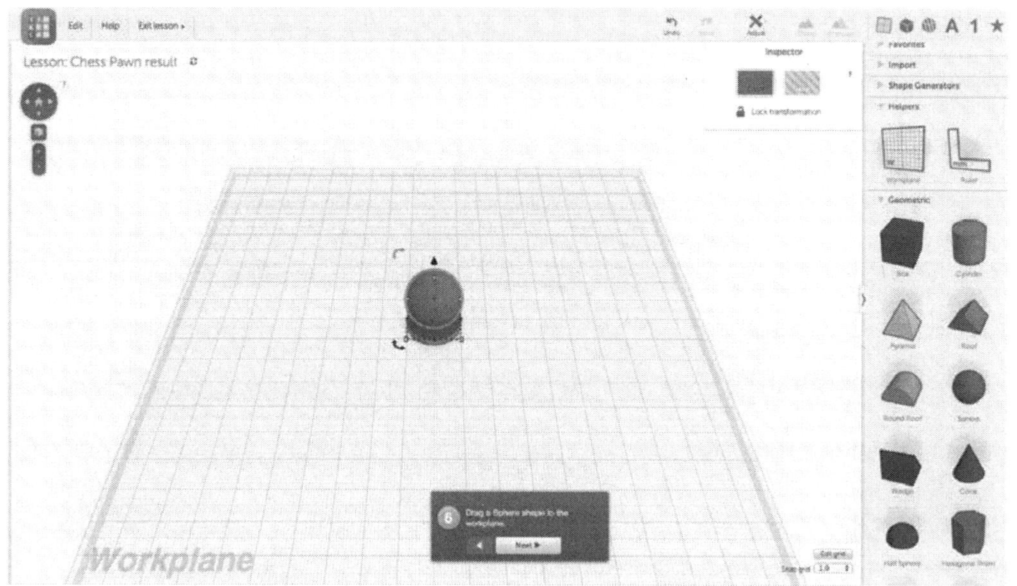

Figure 5-9: Add Sphere on top of Cone

Step 7 - Flatten out the shape of the Sphere and change its diameter. This creates a small ring at the base of the pawn piece.

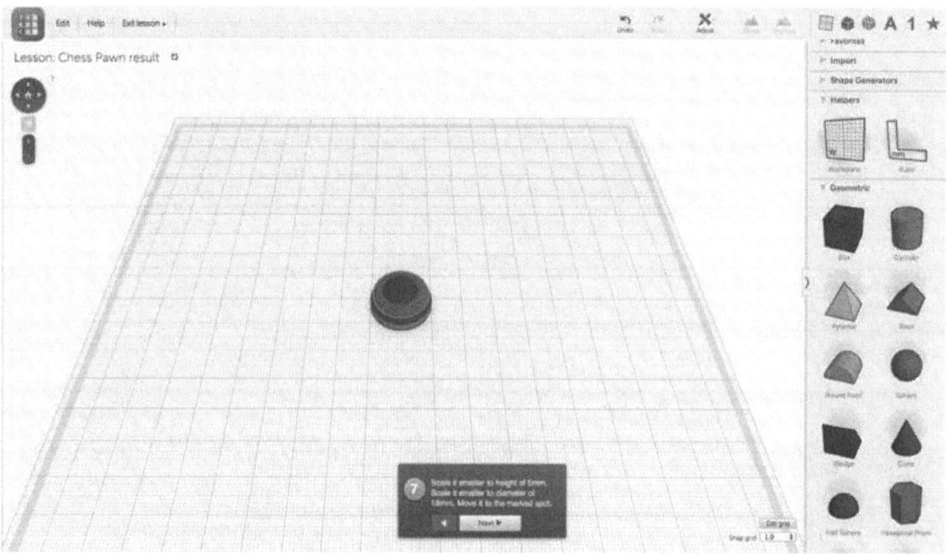

Figure 5-10: Change the shape of the Sphere

Step 8 - Add another cone to the design. This will also be resized.

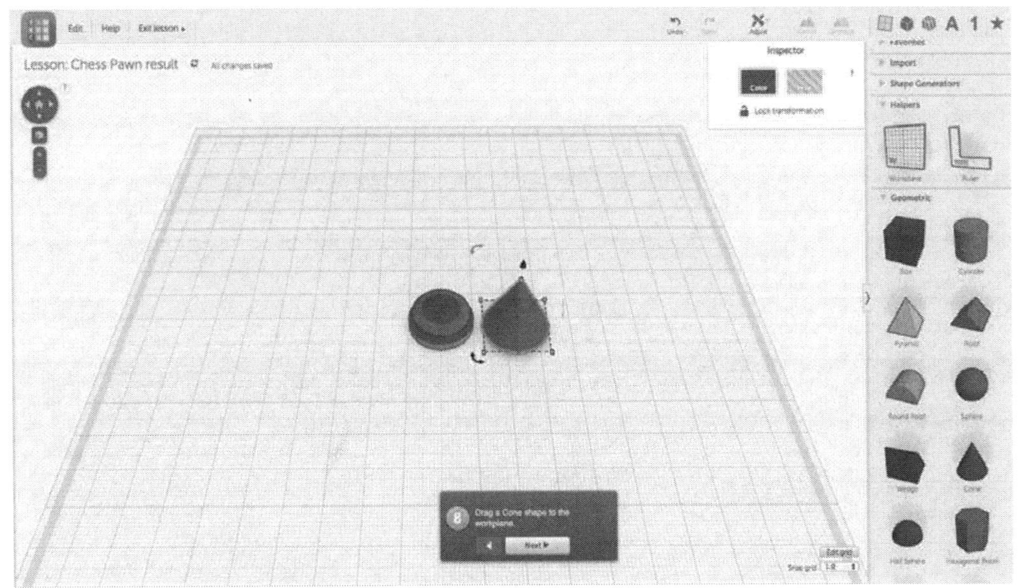

Figure 5-11: Place another Cone in the design.

Step 9 - Change the height and diameter of the cone. Change the diameter first by holding the shift key down and then selecting one of the base blocks and dragging it to the desired size. When you hold down the shift key, all dimensions change together. Then you can change the height separately.

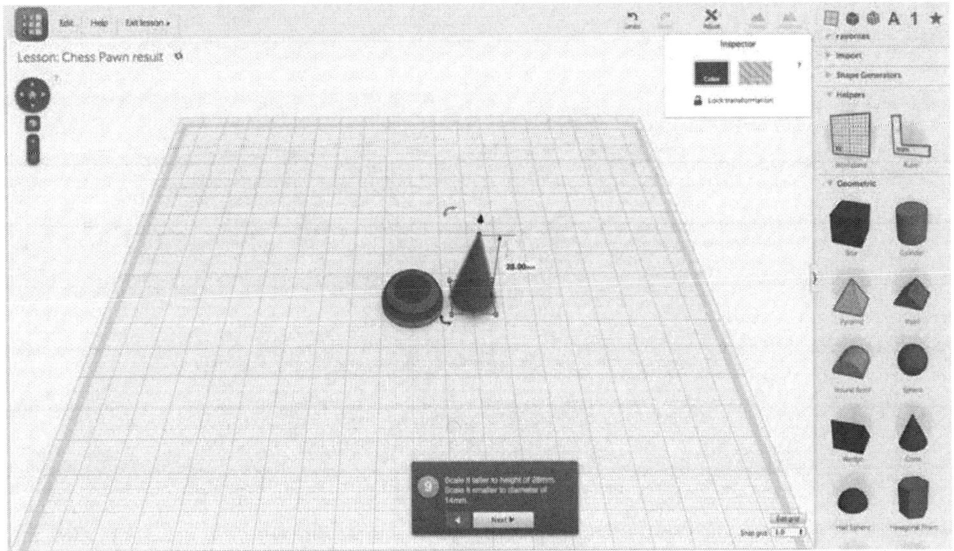

Figure 5-12: Resize the Cone

Pulling up on the top dimension block makes it easy to resize just the height of the cone.

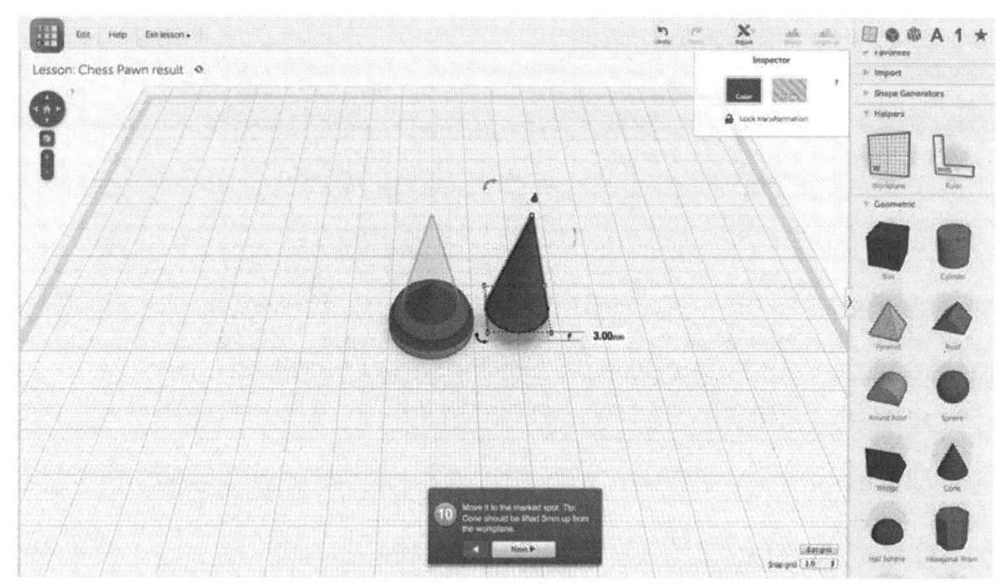

Figure 5-13: Adjust the height of the Cone

Step 10 - Move the cone to the top of the design by moving the cone up with the arrow floating above it. This forms the base of the pawn piece. You can see the design begin to take the shape of a chess piece.

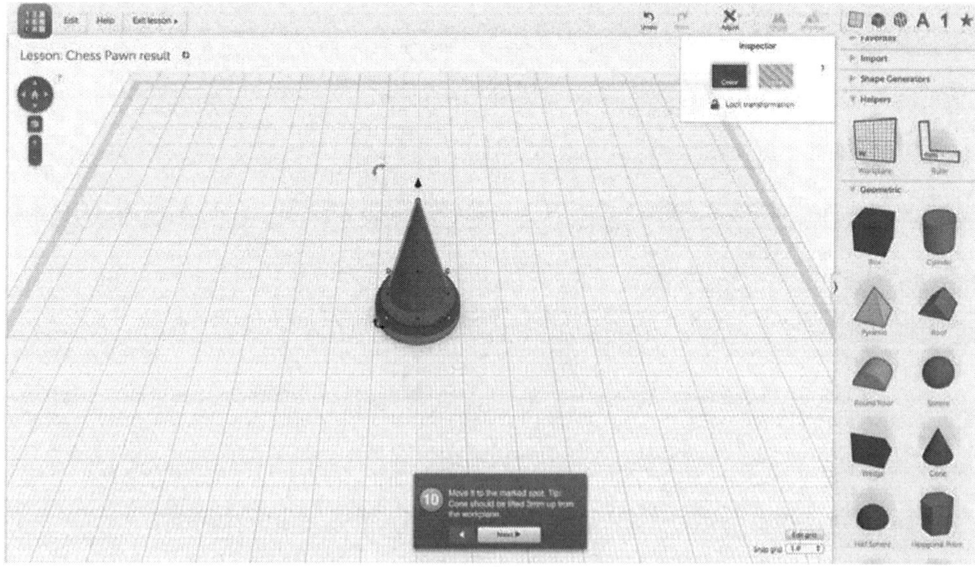

Figure 5-14: Move the Cone in Place

Step 11 - Add another Cone to the design.

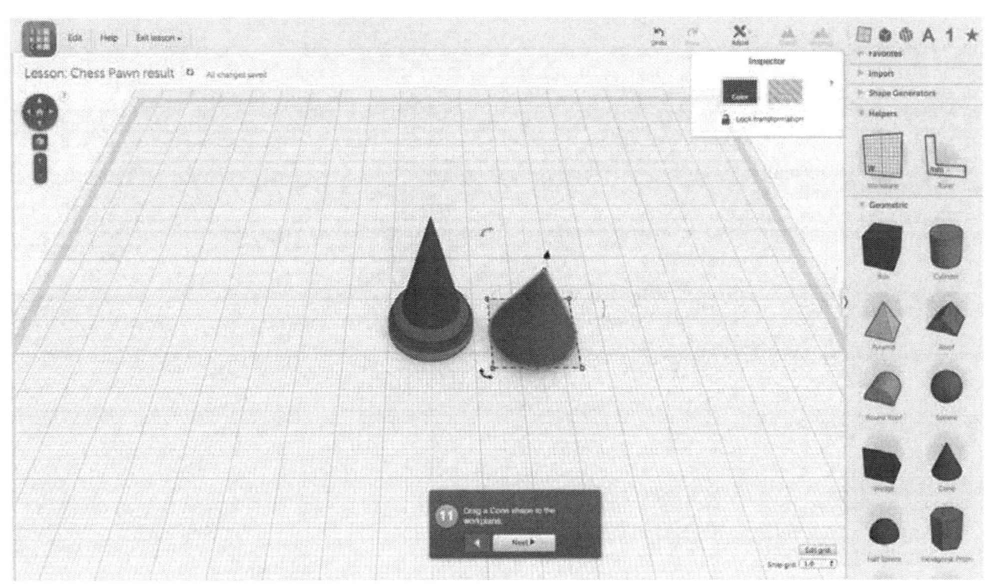

Figure 5-15: Add another Cone to the design

Step 12 - Resize this cone just like the last. With the shift key pressed, drag one of the base squares to resize it to the proper size.

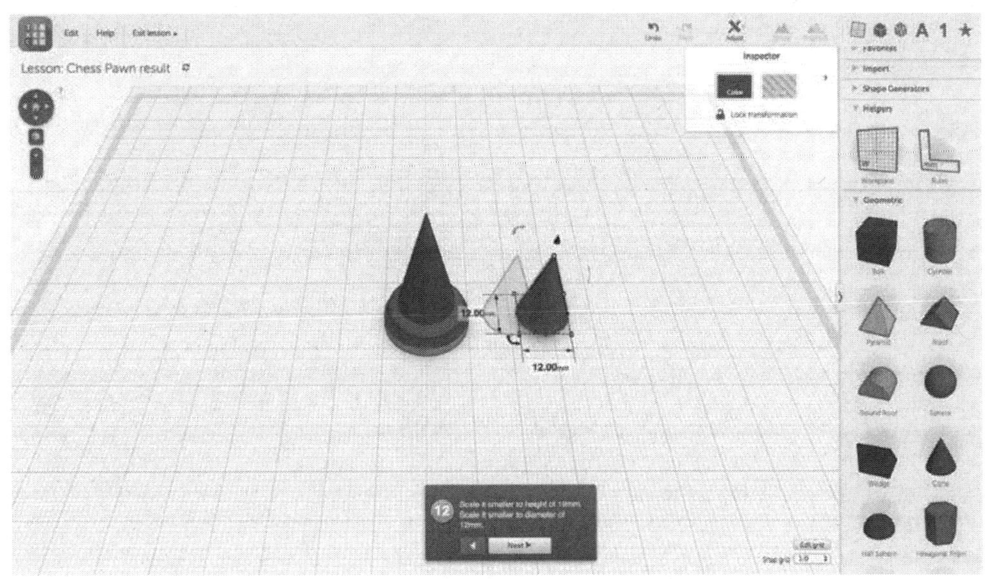

Figure 5-16: Resize the new Cone

The Cone height needs to be set into position for the next step.

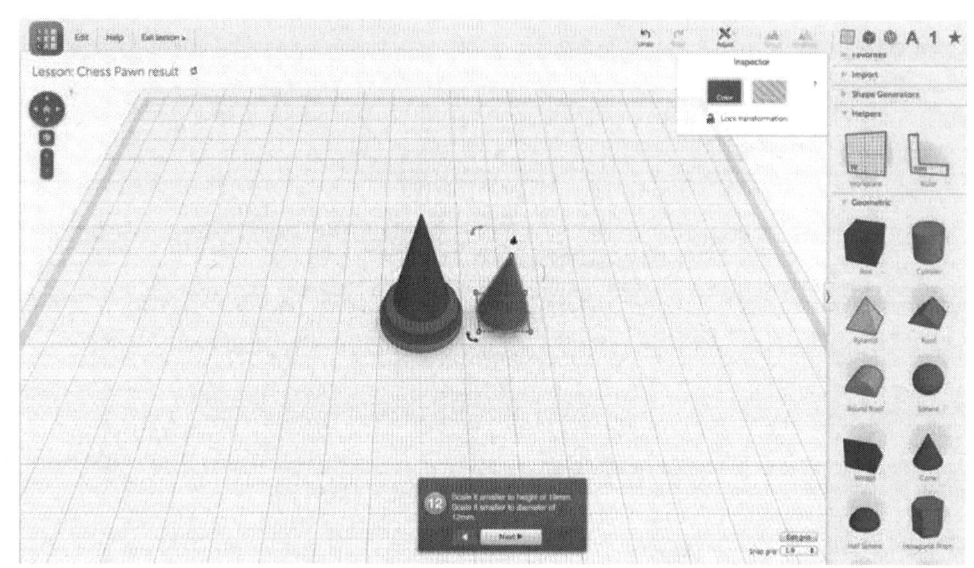

Figure 5-17: Position the Cone

Step 13 - Flip the Cone 180 degrees. This is done by clicking on the quarter round icon with arrows at each end. There is one for each axis so make sure you get the one that rotates it around the horizontal axis. It’s the one near the top of the Cone.

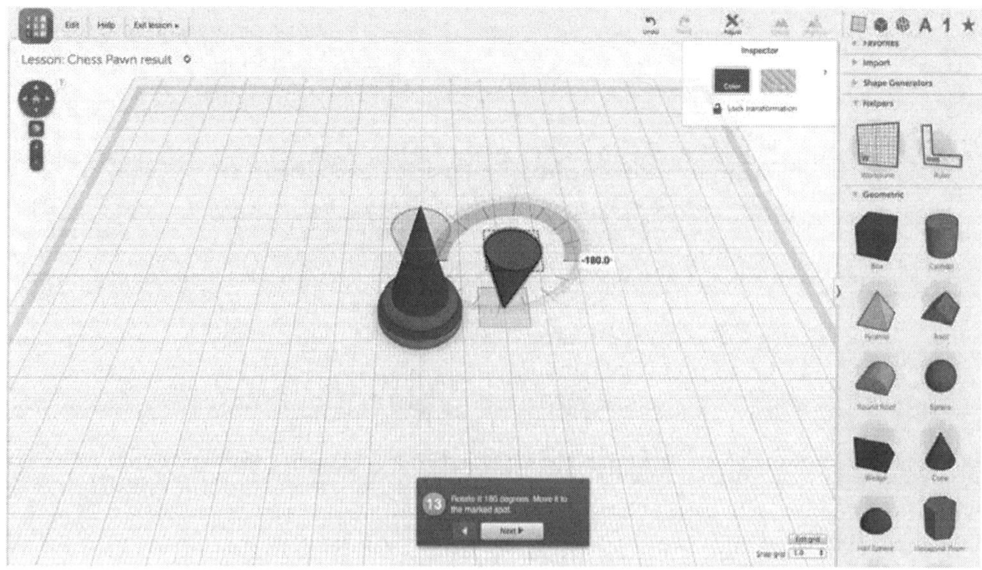

Figure 5-18: Rotate the Cone 180 degrees

Now slide the upside down Cone into the designated area which will merge it with the existing Cone on the pawn base.

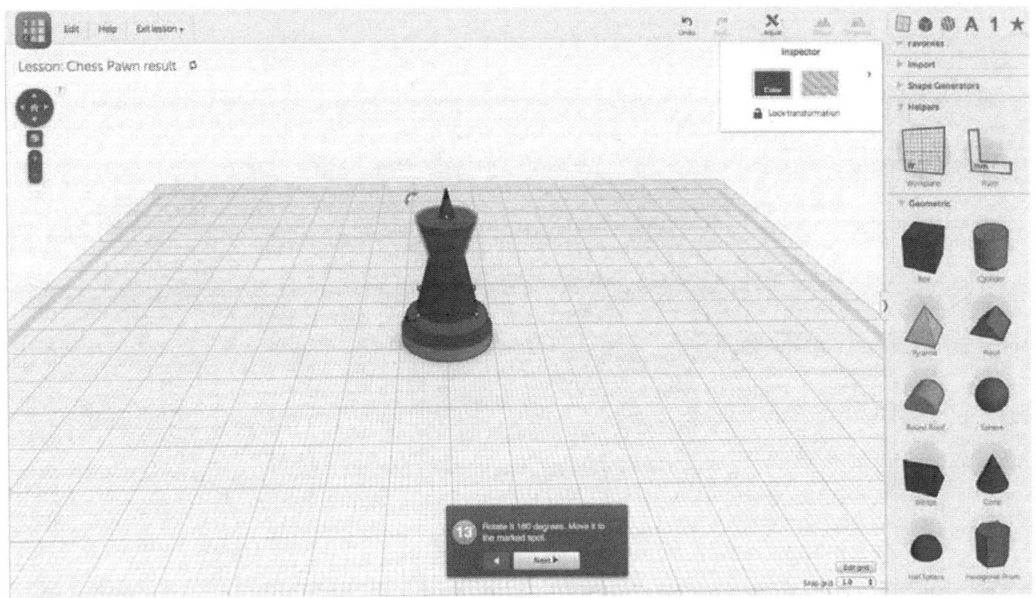

Figure 5-19: Position the upside down Cone

Step 14 - Move the Workplane up higher. This time move it to the top of the upside down cone you just placed. Just grab the Workplane from the Helpers menu and drag it to the flat portion of the upside down cone.

Figure 5-20: Move the Workplane to the top of the upside down Cone

Step 15 - Add a Sphere to the design. Drag it from the menu on the right to the location highlighted.

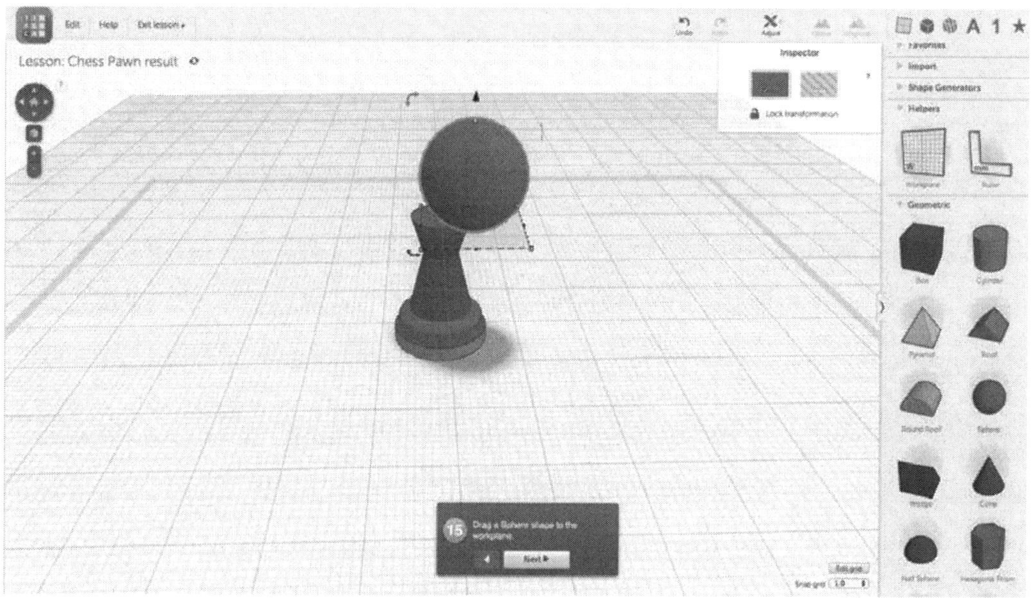

Figure 5-21: Add a Sphere to the design

Step 16 - Resize the Sphere to the dimensions shown and position it in the recommended locator lines. This will place a ball at the top of the Pawn Piece.

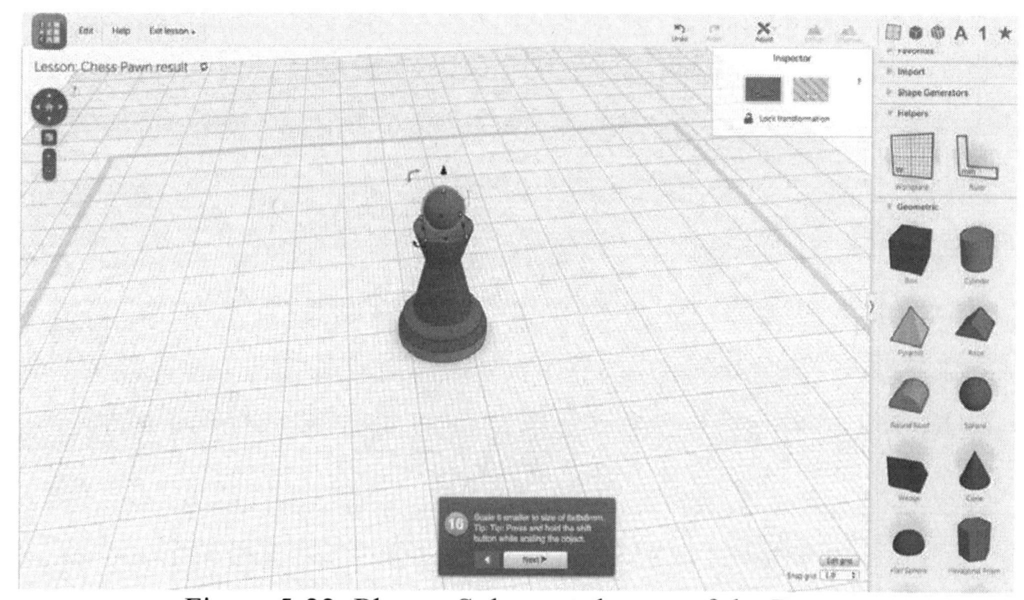

Figure 5-22: Place a Sphere at the top of the Pawn

Step 17 - Return the Workplane to the bottom of the design. Dragging a Workplane image to a space away from the design does this. It will automatically reset the Workplane to the bottom where it originally started.

Figure 5-23: Return Workplane to the bottom of the design

Step 18 - If things don't look lined up, draw a selection box around the whole design so all items are selected. Then go to the upper right menu and click on the Adjust menu and you will see an Align selection. Clicking on that will display the Alignment tool.

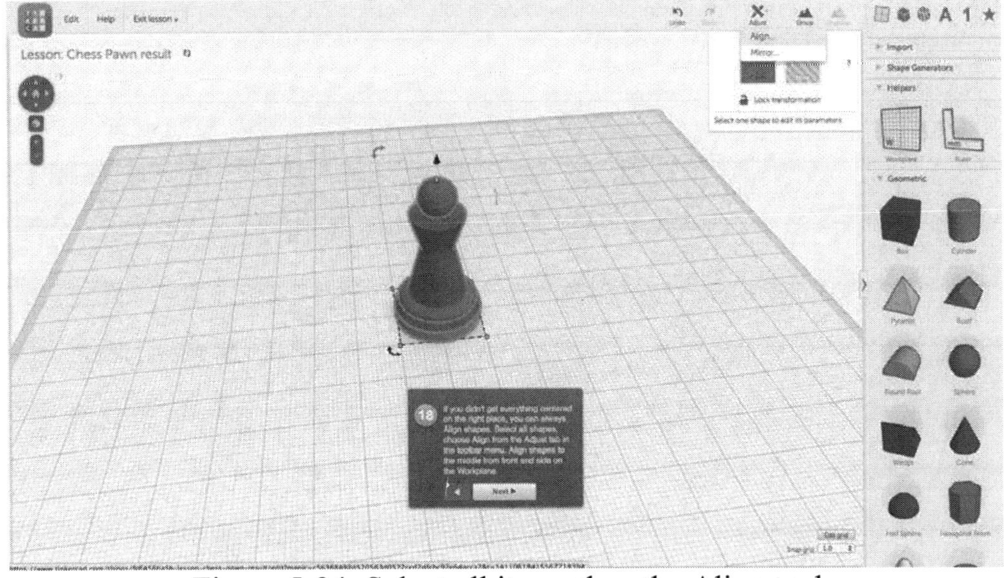

Figure 5-24: Select all items then the Align tool

The Align tool shows dots where the design can be aligned. By clicking on the X direction and Y direction center dots, everything will be aligned at that location. In the picture below, the Y direction locator dot is slightly grayed out and the word Aligned appears, indicating everything is aligned along that axis.

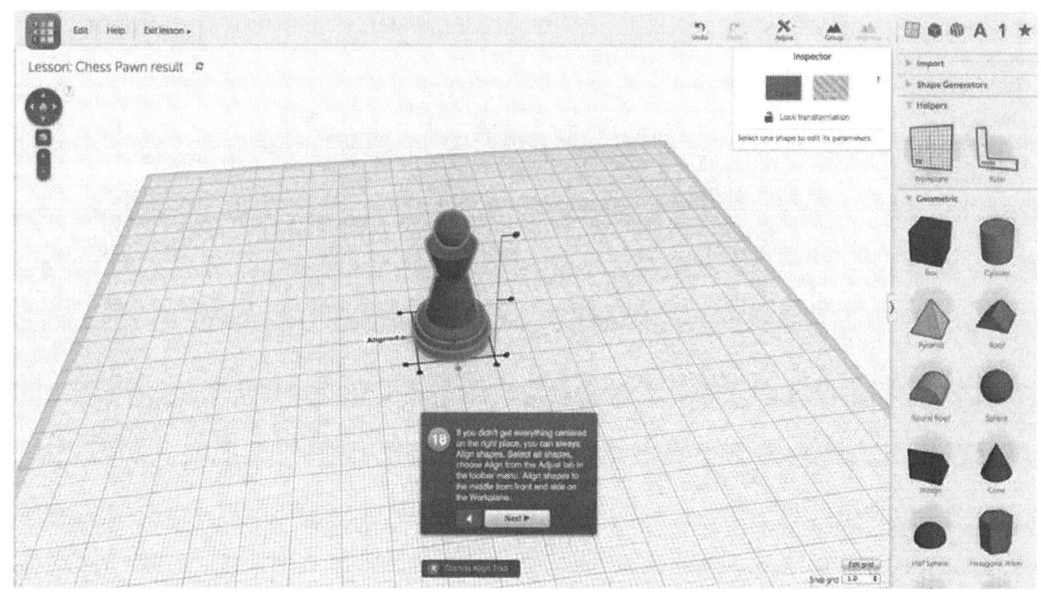

Figure 5-25: Alignment Tool to get everything in order

Step 19 - Now draw a selection box around everything and go to the upper right hand corner and select the Group menu item. This will combine all the different shapes into one solid design.

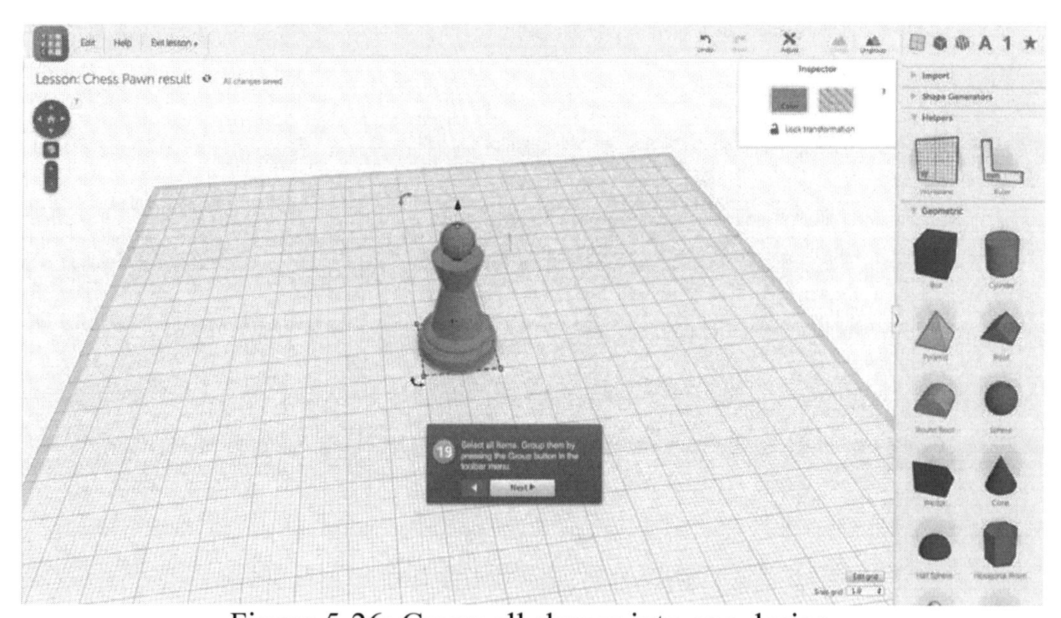

Figure 5-26: Group all shapes into one design

At this point you are done creating the Chess Pawn. Clicking on the Next button will produce a celebration with stars raining down on the design. But you can't print it from here. This is just the end of the lesson.

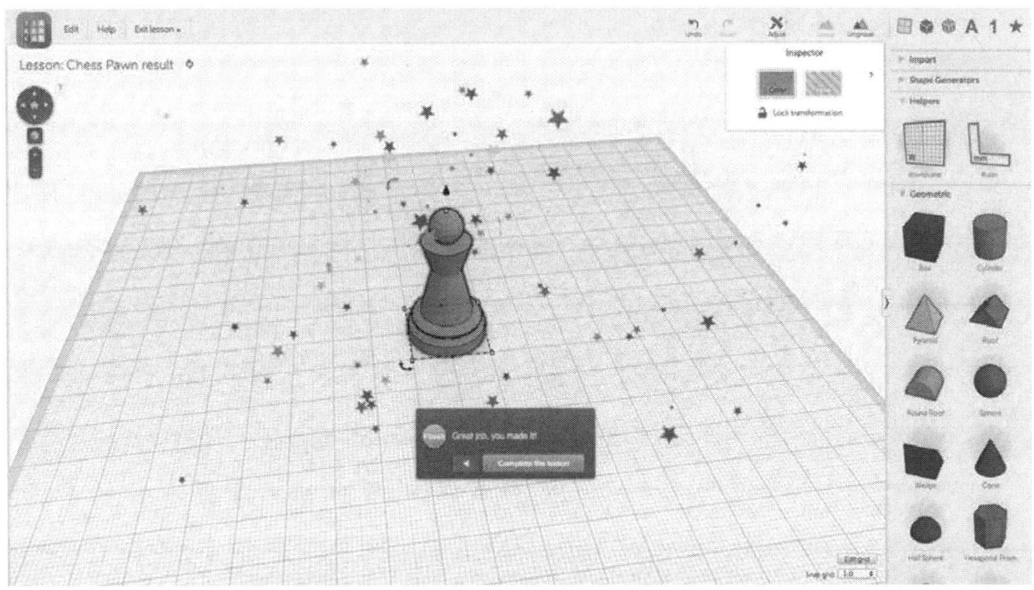

Figure 5-27: Finished Chess Pawn Design

While you are at this screen click on the Edit menu and click Copy. This will copy the design to the clipboard (make sure the design is still selected by drawing a box around the design before hitting copy).

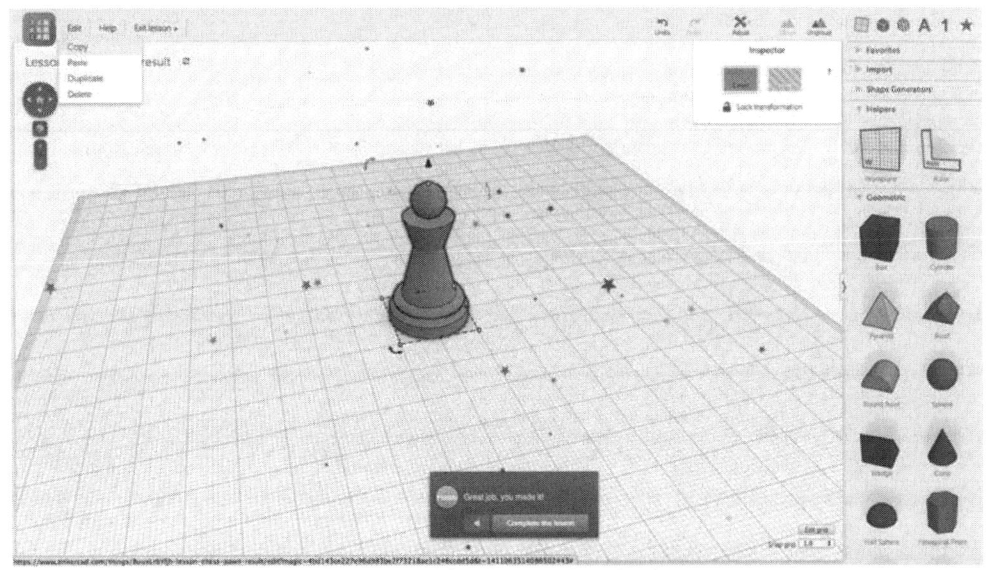

Figure 5-28: Copy the design

Then click on the Tinkercad logo in the upper left corner and you will go back to the main Tinkercad screen. Here you click on the "Create new design" button to bring up a new blank work area.

All designs

Here are the designs you have created or which have been shared with you.

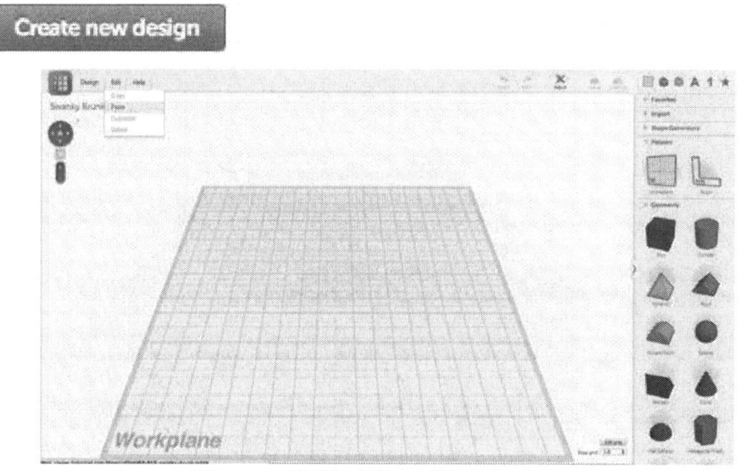

Figure 5-29: New Workplane

Click on the Edit menu and then Paste to create a copy of the Chess Pawn Design. It will appear on the platform but will have a crazy made up name in the upper left hand corner. In Figure 5-30 its shows the name Swanky Krunk.

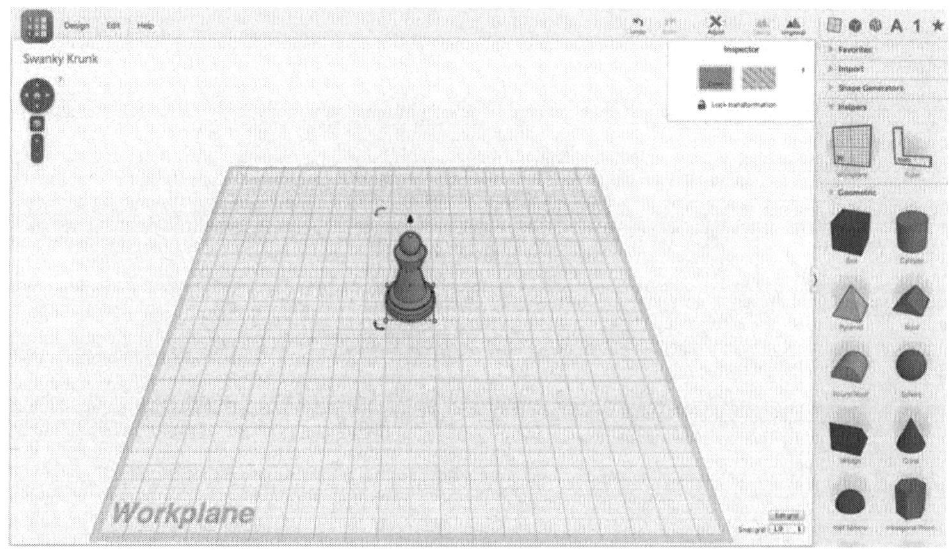

Figure 5-30: Pasted Design with new name

Click on the Design menu and select Properties. This is where you can change the name to anything you want.

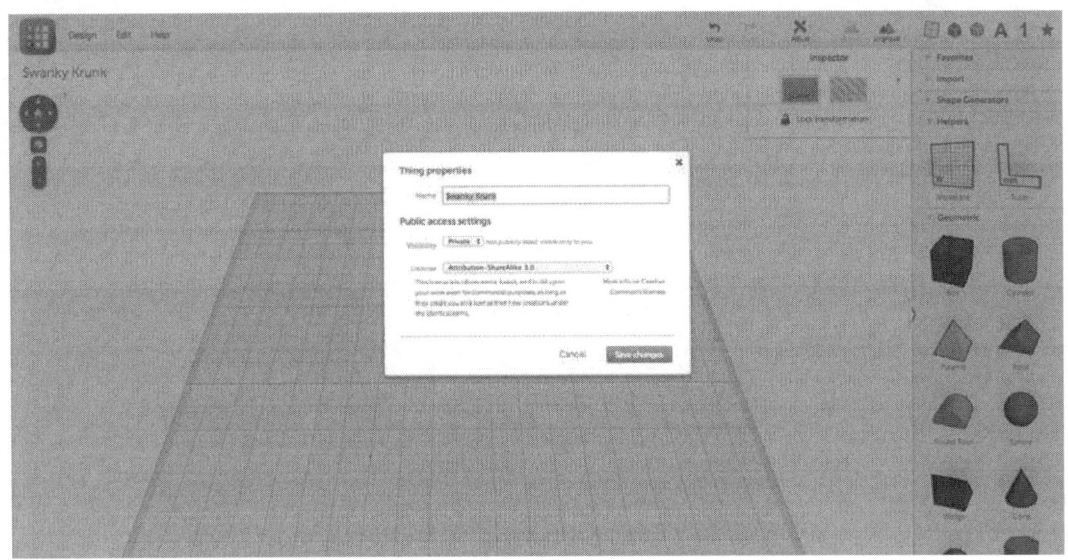

Figure 5-31: Properties window

You can also make the design public here so anyone can use the design you created.

Figure 5-32: Make design Public

Now we are ready to create the .STL file for printing. Click on the Design menu again only this time select Download for 3D Printing.

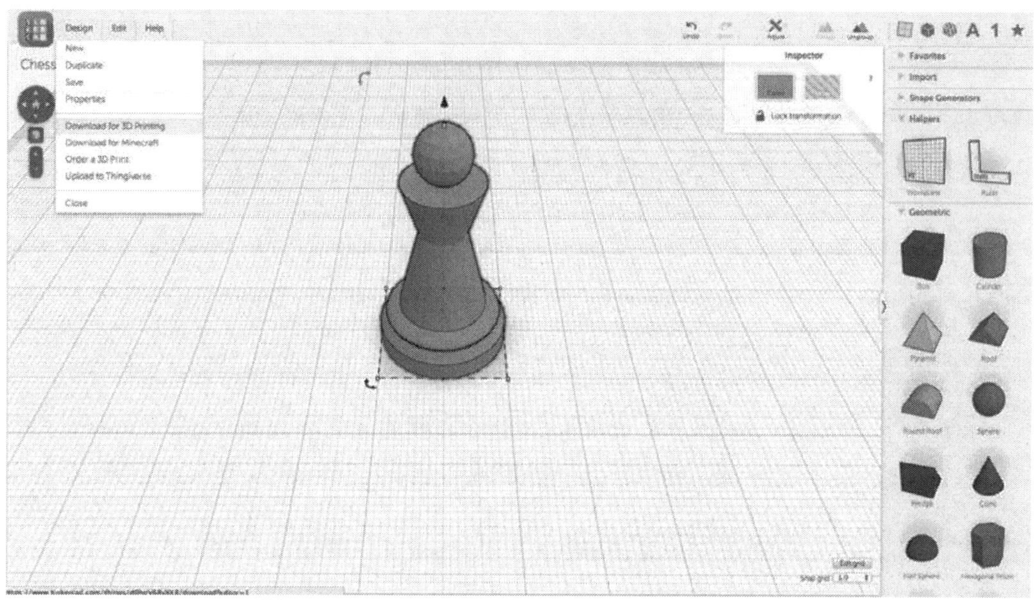

Figure 5-33: Download for 3D Printing

We select the .STL file option from the pop-up menu.

Figure 5-34: STL menu selection

The .STL file will appear in the lower left corner of the screen and will have already downloaded to your computer through the Chrome browser download utility. It will have the name you gave it in properties dialog box followed by the .STL suffix.

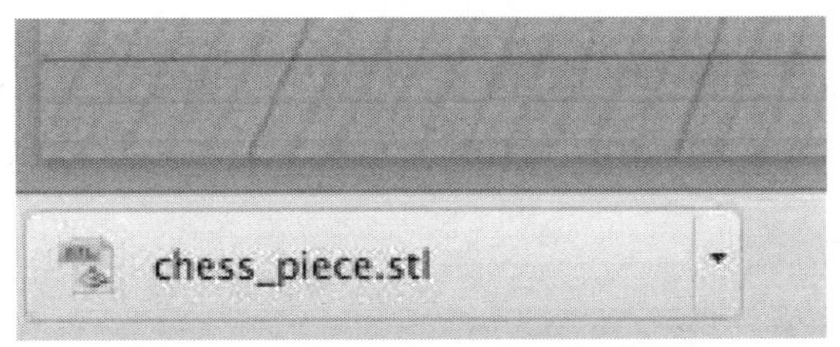

Figure 5-35: Downloaded File

From here we are actually done with Tinkercad. We designed a custom creation using various different shapes. Then we grouped them together and saved them as one design. From there we had Tinkercad create a .STL file so the next step is to load it into the XYZware that the Da Vinci printer uses to load the design.

This simple design should demonstrate that with a little creativity you can create all kinds of designs with Tinkercad. One thing this didn't show is that any shape can be changed to a hole. A hole will take away material instead of adding it. We could have placed another cylinder inside the chess piece and made it smaller than the main body. Then we could change that smaller cylinder into a hole to make the chess pawn hollow. If you want to learn more about this, I suggest you run the Wrench tutorial where you create a plastic wrench.

For now, lets print the Chess Pawn piece we created. Open the XYZware and load the .STL file.

XYZware

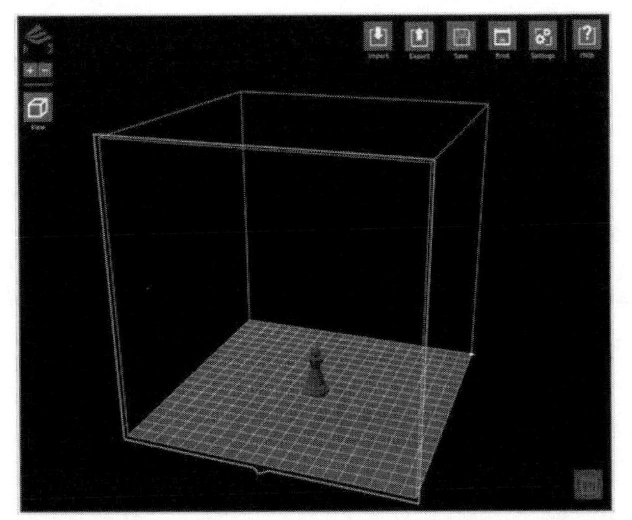
Figure 5-36: Chess Piece loaded into XYZware

We can click on Print to slice the design into the G-Code file and then it will be ready to send to the Da Vinci 1.0 printer.

Figure 5-37: Slicing of the .STL file

The design is ready and shows it will take 0.93 meters of the 7.53 meters left on the cartridge. It estimates it will take 30 minutes to print.

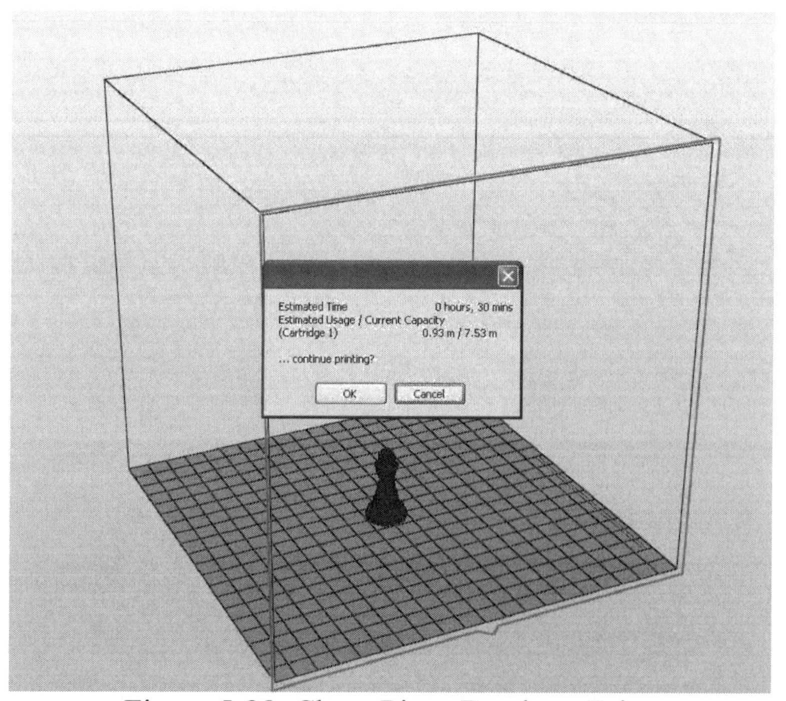

Figure 5-38: Chess Piece Ready to Print

The design is sent to the Da Vinci 1.0 printer and the printing begins.

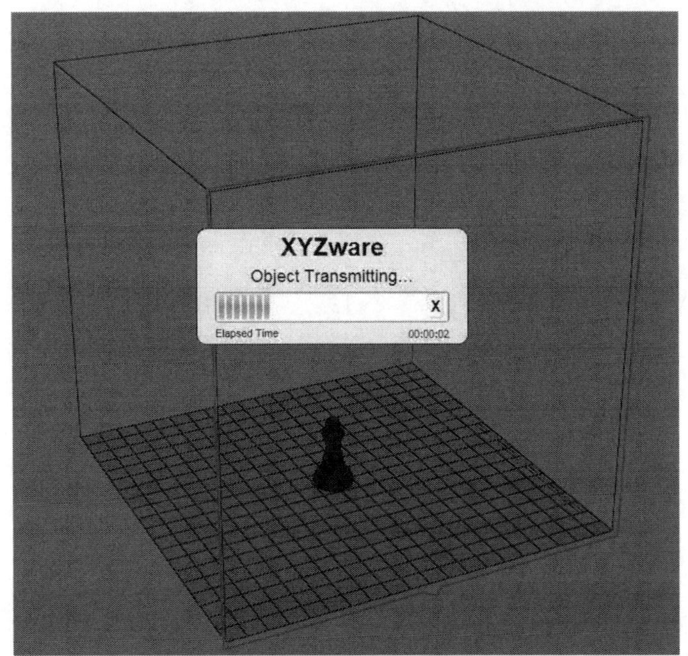

Figure 5-39: Design being sent to Da Vinci 1.0 printer

3D Print

This print turned out to be a great example of issues that can occur with some designs. The ball at the top of the pawn was set to 8-mm diameter but the Da Vinci had a problem printing something that small. The head moved so fast that it would not give the plastic enough time to harden. A simple fix is to make the ball on top 10-mm diameter. This gives the plastic time to harden.

You can slow down the print head, as that is one of the XYZware settings, but I tried that without success. After you print a few designs you'll start to learn what you need to adjust for quality prints. I also enabled supports since the top half of the pawn was sloping outward. Supports add plastic to support the layers hanging out and then the supports can be easily broken away.

On the left in the picture below is the 8-mm ball and on the right is the 10-mm ball. You can see what a difference a size adjustment can make.

Figure 5-40: 8mm vs 10mm ball Pawn Print

Another trick is to print more than one pawn piece at a time. That way the extruder head has to move between each pawn piece on every level. This takes more time since the head is back and forth so much, but it does give the plastic time to harden in between layers. You probably would want more than one pawn anyway so it's worth considering. This idea of printing more than one can be very helpful. When I need to print a design with small features on top, I often print two and spread them out to give the plastic time to cool between layers.

Chapter 6 – Scanning a 2D Object

Another feature of Tinkercad is the ability to import a two-dimensional drawing as a .SVG file and then use the Tinkercad features to make it a 3D object. That means you can scan in a drawing or the flat side of an object and make a 3D version of it. For example, I had a broken vertical blind valence clip that was difficult to find so I placed a good one on my paper scanner, then converted the scanned image into a .SVG file format and loaded it into Tinkercad. From there I adjusted the height to make it a 3D object and then adjusted the length and width to match. The final design was sent to my Da Vinci 1.0 printer and the result is shown below. The original is in white and the 3D printed version is in black.

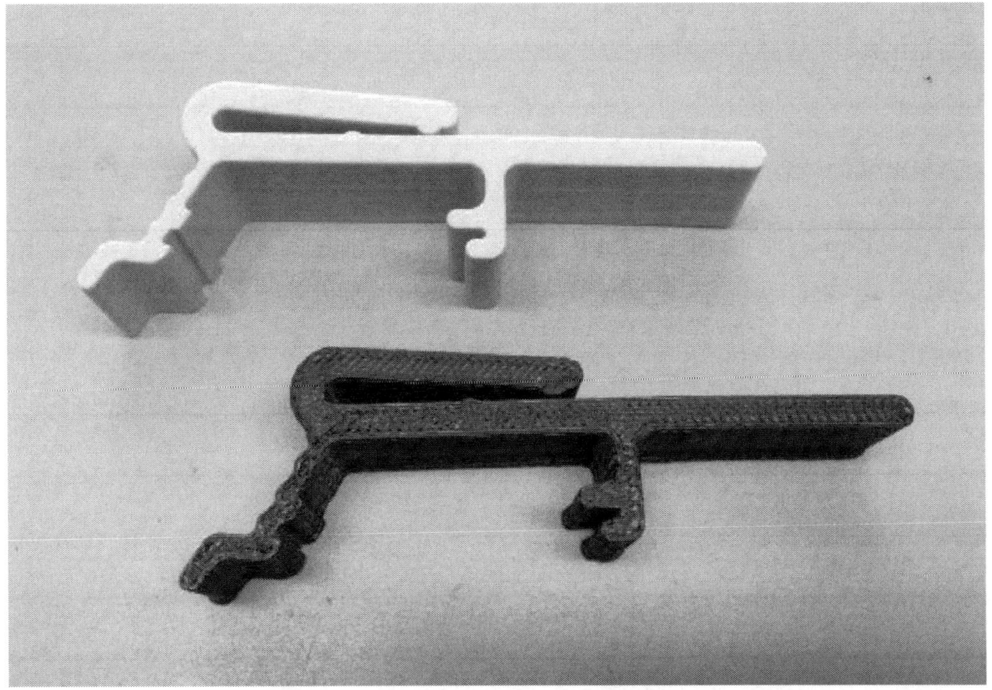

Figure 6-1: 3D Printed Valence Clip

This is really a great way to create prints if you are new to using CAD software. As easy as Tinkercad is you can't get much easier than scanning an object like it's a document.

I'll step you through a scan of my name and then 3D print it. You can then do the same for your name.

Printing Your Name

I took a black marker and wrote my name on an 8 ½ x 11 white piece of paper. Then I scanned that document into my computer as a .PNG image. My computer wouldn't allow me a direct scan to .SVG and I suspect most computers won't offer that. The scanned image is shown in Figure 6-2.

Figure 6-2: Scanned Image of Name

The scan will most likely have a white or grey background but that will count as a layer when converting to .SVG for 3D printing. So we have to make it a transparent image and eliminate the background layer. I used the Instant Alpha tool on my iMac's Preview application. It allows me to outline the name and delete the white surrounding the name. If you are good with Printshop or other image editing software, then you may have an easier way to do it. The result is the image in Figure 6-3. It probably doesn't look much different than Figure 6-2 but it actually has no color at all around the black marker writing. We need a very simple image to convert to .SVG and only the parts we want to print need to be present.

Figure 6-3: Transparent Version of Name

In order to import it into Tinkercad, I needed to convert the file from .PNG to .SVG and luckily there are many online conversion services for free. I used the one at:

image.online-convert.com/convert-to-svg

I'm sure there are many other options and many programs you may have that can make the conversion. I tried to use a free program called Inkscape but didn't have a lot of luck with it. The image was not clean and I found Inkscape a bit difficult to work with even though I've used it for CNC router projects before.

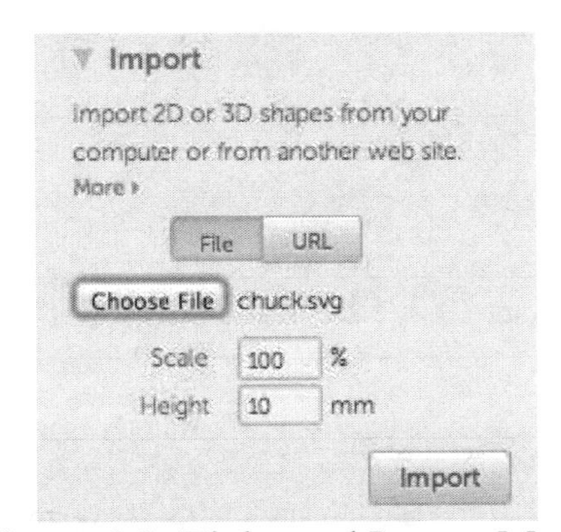

Figure 6-3: Tinkercad Import Menu

I then imported the .SVG file into Tinkercad from the import menu. You can set the scale and the height before clicking on import. You will probably have to downsize the file a lot. I imported it at 20% and it still came out large compared to the platform as you can see in Figure 6-4. I left the height at the default 10 mm. That seemed tall enough for this build.

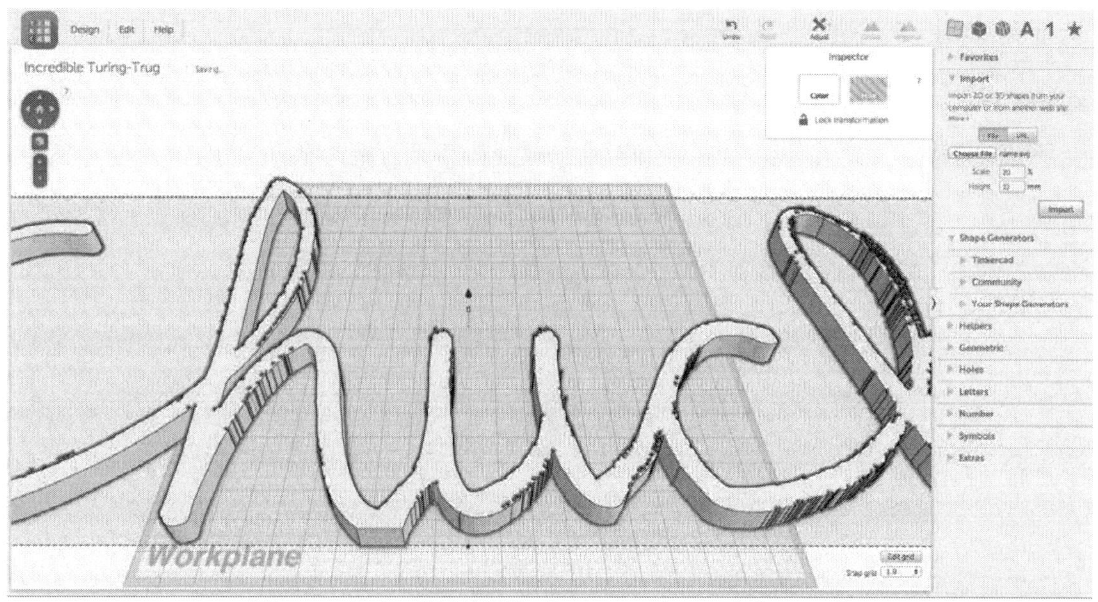

Figure 6-4: .SVG File Imported to Tinkercad

I resized the 3D design to fit on the Da Vinci 1.0 printer bed which is about the same size as the default Tinkercad base. The result is shown in Figure 6-5. The best way to do that is to grab a corner with mouse and then hold down the alt key as you move to resize it. This will resize everything in the X and Y direction proportionately but leave the height in the Z direction alone.

Figure 6-5: Resized Name

The file was then exported as a .STL file the same way we exported the chess piece in the previous chapter. Then the .STL file was loaded into XYZware, sliced and sent to the printer.

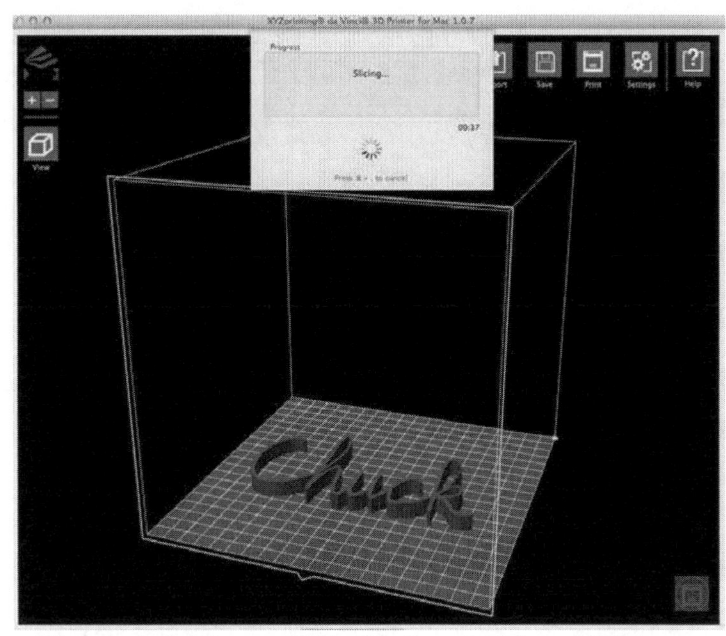

Figure 6-6: Name .STL file Loaded into XYZware

I didn't need high resolution so I printed it at a 0.4 mm height and 10% fill. The result was quite good.

Figure 6-7: Finished 2D to 3D Print

Now this doesn't have to be a scan and print design on its own. You can draw out a design and then incorporate that design into another design. I've seen people make some really interesting patterns on paper, scan them in and then modify the back of an iPhone case design so they have a custom "one-of-a-kind" case for their phone.

Chapter 7 – Sending a Design to a Professional 3D Printer

Tinkercad offers the option to send your 3D file to a professional 3D print company. They are linked so it's a simple process of selecting your choice and then Tinkercad will upload the 3D file.

Ordering a 3D Print

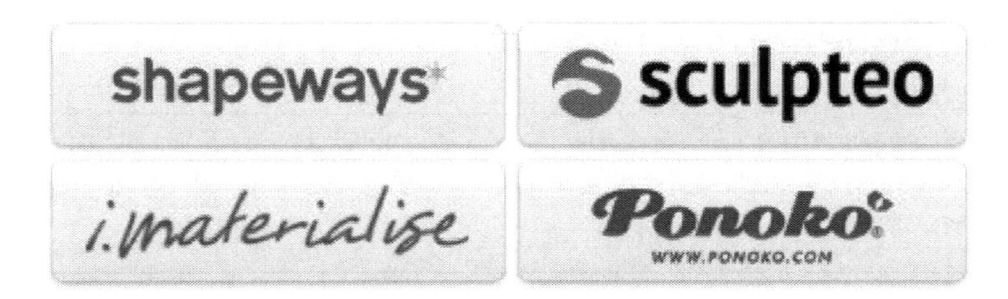

Figure 7-1: Professional 3 Print Suppliers

I chose Shapeways as they are one of the biggest. I sent the pawn piece we created a couple chapters ago and the abbreviated results are shown in Figure 7-2. Strong and Flexible plastic would cost $7.15 for one piece in white. Black, White Polished and Purple Polished would cost more.

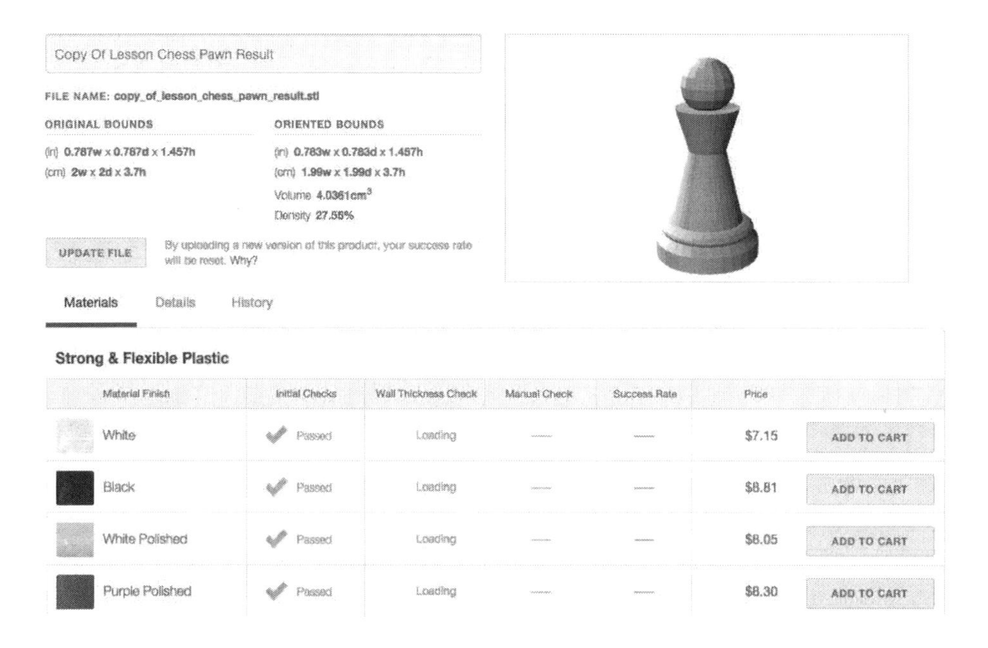

Figure 7-2: A Few Print Options

These aren't the only options though. There were 12 colors to choose from plus you could get it printed in Metallic Plastic, Detail Acrylic, Stainless Steel, Precious Metal (Silver, Gold Plated Brass, etc.), Sandstone and Wax. Some designs can also be printed in ceramic but this piece was too small.

The cost of the other services are much greater.
Metallic Plastic ~ $10
Detail Acrylic ~ $15
Stainless Steel ~ $40
Precious Metal ~ $75 to $7,000 (platinum)
Sandstone ~ $6
Wax ~ $42

Sandstone seemed the most interesting to me since is really sedimentary rock consisting of sand or quartz grains cemented together. Seemed like more of a "chess piece" style and was the cheapest of the bunch.

I decided to order one to see the results. The order process was easy enough. I just had to create a login and then give them my payment and shipping information. There were several different shipping alternatives. I chose the cheapest, which was USPS.com First Class Shipping. The total cost including $4.99 shipping was $11.02 for one chess piece. Since every piece is built individually, and not from a mold, the cost of multiple pieces doesn't reduce the price. If I ordered three pieces then the cost would triple from $6.03

each to $18.09 for three. You would save on shipping but not on the print.

Figure 7-3: Volume Pricing Comparison

The piece took a little more than a full week to arrive. You can see it in Figure 7-4. This is a great option. I can print out the design on my Da Vinci and if I want or need a professionally printed part, I can easily get that delivered in a relatively short time frame.

Figure 7-4: Sandstone Chess Piece from Shapeways

Chapter 8 - Da Vinci 1.0 Printer Tips and Tricks

The Da Vinci is a great 3D printer for the money and with software updates it continues to get better but there are some things that you can do to make it better or get better prints out of it. I will cover some of the techniques I've discovered from using the printer for a while.

Clean the Extruder

Before every print, make sure the extruder is clean. There is always plastic left on the head after a print and if it starts a new print with that plastic in place, it will get into your new prints. Sometimes this can cause a print to bunch up on the bottom of the extruder and you end up with a glob of plastic. You can use the wire brush that came with your Da Vinci to clean the extruder.

Printing in a Circle

If you have one item to print, place it in the middle of the bed. This will give it equal shift left or right for any tolerances that may be in the Da Vinci mechanicals. But when you have to print more than one item, place them in a circle. The extruder head can sometimes cross over one print on its way to another print. I've see the hot extruder grab a plastic piece as it moved and then that plastic gets molded into other pieces, messing up the whole print.

If you place the items in a circular pattern, then the extruder will move from one to the next without crossing in most cases. I've had great success with this method and I highly recommend it.

Glue

The Da Vinci comes with a glue stick for holding down prints. Getting that first layer to stick and hold is critical to getting successful prints. I've found that too much glue can affect the heat coming through the glass bed, but putting at least some glue down can be very helpful. Also, cleaning the bed from previous glue is very important. Old glue that is heated and dried ends up making an ugly residue on the bottom of your prints.

To remove glue, take a washcloth and get it wet with warm water. Wring it out so it's wet but not dripping. Place the wet cloth on the base and let it set for a few minutes. This will soften the existing glue and then you can use the cloth to wipe off any existing glue easily. Then dry the base with a dry washcloth or paper towel. After it's dry add glue to just the area where your print will land.

Here is a big tip: also place glue where the test strip will run. This may seem dumb but if the test strip doesn't stick it will get dragged over into your first print and sometimes dragged into the second print. It can also cause a bunch up on the extruder head and cause all kind of problems. So making the test strip stick is just as important as the main print.

Some people use Kapton tape or sticky blue tape but I've found the glue works fine. Make sure you use washable stick glue.

Raft

Printing a raft first can definitely help make the print stick to the bed but removing the raft is an absolute pain. You end up scraping, cutting or sanding off the bottom of your prints. Only use a raft when the bottom of your print isn't that important to be completely flat and you want that extra insurance that the print will stick to the bed.

Low Spool

When your spool is low, don't print a large print. I've found that sometimes the length of the remaining cartridge is wrong and the spool runs out before the print is done. XYZware will tell you if there isn't enough plastic to print the design, but if the number is wrong it will keep trying to print while the plastic has run out. This will leave your extruder clogged and it is a royal pain to get it out and cleared. So save the low spool sizes for small prints. You can also take apart your cartridge and measure the plastic left just to be safe.

Heated Bed Adjustment

One of the biggest improvements I found with the Da Vinci was the heated bed adjustment. The software has an automatic calibration routine you can run, but all it does is check three corners of the heated bed and then gives you a value for each corner. If the value is between preset values for all three corners then the Da Vinci will report that the bed is leveled. If the numbers are outside the range, then you need to turn the three spring loaded adjustment knobs to get the bed into the range. This method, although rather interesting, has issues.

First off, the checkpoints are at the corner of the bed not on top of the adjustment screws. Therefore, if one corner is out of adjustment, you may need to balance the adjustment between two knobs and that can therefore affect another corner that may have been within range.

I found that manually adjusting the bed while a print is first starting can be the best method to get a properly adjust bed. In fact some of my best prints have come from a bed adjusted to a level that would never pass the auto-calibration method. Thankfully the Da Vinci will still print your design even if it thinks the bed is out of adjustment.

The trick is to set a starting point and then adjust from there. You could manually move the extruder around the bed and adjust the screws so the nozzle is the same height across the whole print bed but there is one problem with this: you don't know the starting level of the bed height. The Da Vinci seems to constantly adjust the bed just prior to actually laying down the plastic. For this reason some people use a piece of paper as a measuring tool. You start a print and then wait for the bed to get in position and the extruder to begin laying down the test strip on the edge of the print bed. Then mid-way through the test strip, shut off power to the printer. This leaves the bed at the proper height and you can move the extruder around. To do this though, you have to wait for the extruder and heated bed to cool down.

Once everything is cooled down, place a single sheet of white printer paper on the bed and then manually move the extruder around. Start by positioning it above each adjustment screw and adjust it so the extruder is just barely touching the paper. You can slide the paper around to test how much clearance there is. You should be able to just barely move the paper.

Once the adjustments are made, then it's time to start a print again. I suggest making a simple print of a small circular object such as the chess piece or just make a base of the chess piece as the test print. Once you have that created, then place the object in the XYZware and place it near the front of the base. Then place the same object seven more times so they form a circle around the base. Now run the print and watch the test strip print closely. This is the long strip that prints on the right side of the base at the beginning of any print.

The first layer is key for all eight pieces. The first layer should be semi-flat as it prints so it holds to the glass base but also not too flat that the extruder hits the glass and just leaves blobs around the extruder nozzle. As the extruder places the test strip, watch it closely. If it looks like it's not being squished then adjust the knob in the right rear to get the test strip to flatten out. You may also have to adjust the front knob, but get the right rear knob adjusted first.

Next, watch where the eight prints start; chances are the first one will start in the front near the front adjustment knob. Be ready to adjust the front knob to flatten out the print. If it's too flat or no plastic is coming out, it's too close to the base and needs to be adjusted down. If the print is not flat at all, then adjust the base up. Finally wait until the prints on the left side of the base begin to print. Adjust the left rear knob to flatten the prints properly. Once you have it adjusted you can let the print finish and check how they print across all the eight pieces. There is a very good chance that making these adjustments while printing messed up the print. That's OK. Cancel the print, clean off the base and then start the print again.

I've actually measured the thickness of different adjustments and now I can tell the proper setting by looking at the width of the first layer. Figure 8-1 shows how various thicknesses look. I manually adjust the bed while that first layer is going down to achieve the width for the 0.25mm thickness.

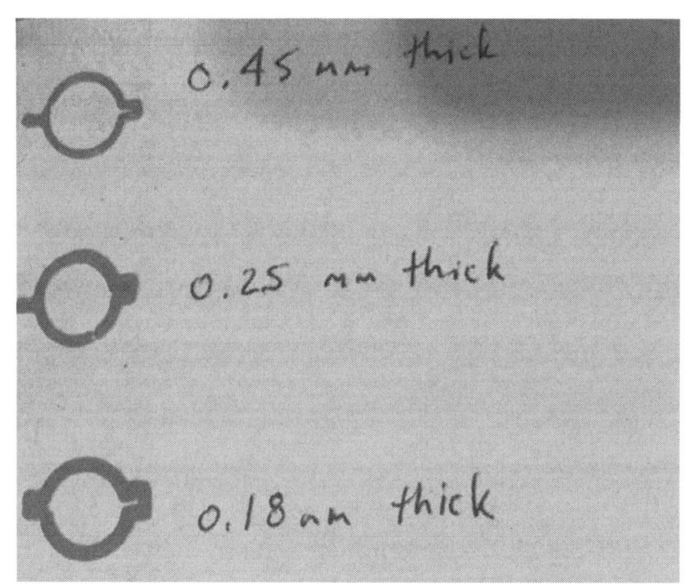

Figure 8-1: Thickness of First Layer

Once you get your bed adjusted for that first layer thickness you want, you should be able to get very accurate prints that stick to the glass base without any glue, though I still recommend a slight amount of glue.

I have a pattern of twelve pieces I use to adjust the base. They had critical little tabs that I needed for another project and getting these right took some time. So now I use them as the test for adjusting the base. If I can get these to print accurately, then any other print I run come out perfectly.

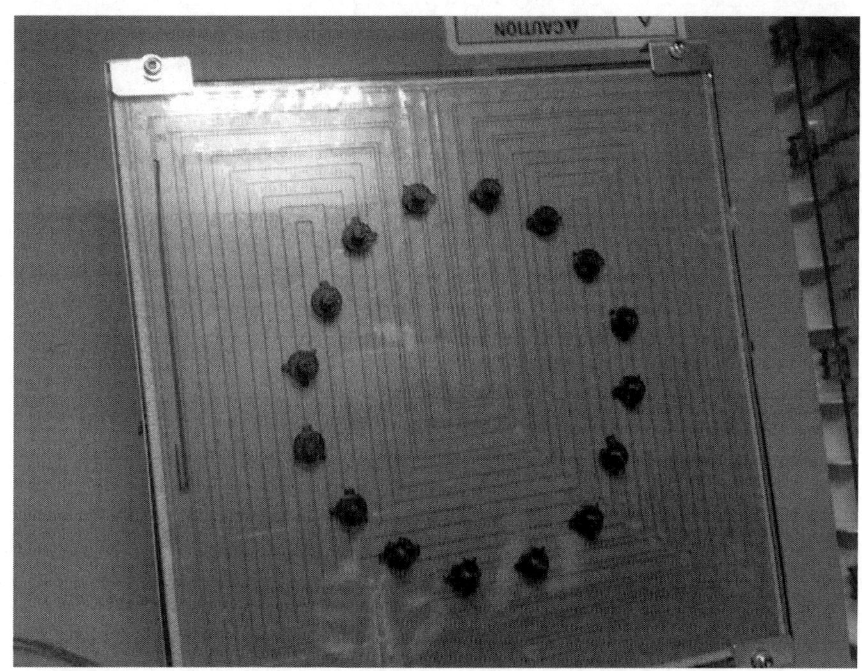

Figure 8-2: Circle of Objects for Heated Bed Adjustment

Da Vinci Maintenance

I've recorded some maintenance tips that I've made to help make my Da Vinci print a whole lot better. They are not adjustments I necessarily recommend for the beginner but if you are ready to tackle them, here you go. I also have these shown on my YouTube Channel at:

www.youtube.com/user/beginnerelectronics

Belts

The belts in the Da Vinci should be tight. Not so tight that the shafts can't turn easily, but tight enough so there is no slippage. Each belt has an adjustment but requires the plastic sides to be removed from the Da Vinci case. They just snap off at the top and lift out so they are easy to remove and replace.

The X-Axis belts have spring-loaded adjustments so just loosening the shaft bearing mount screws will allow the springs to tighten the belts. Then just tighten the bearing mount screws.

The Y-Axis belt is a little tougher. The adjustment is held in place with one screw that is accessed from the back of the printer. You have to reach inside with a Phillips screwdriver to loosen the screw and then pull back on the adjustment with your other hand while you tighten the screw.

The Z-Axis has a screw drive so there is no belt to adjust.

Base Wobble

The belts affect the X and Y accuracy but even more critical is the wobble in the base. The glass on the base can become loose on some printers. This is a problem because as the extruder is placing plastic on the print, it is essentially pulling the print back and forth. If the glass is loose, the X and Y direction can be affected. On one printer I saw a whole shift of layers due to the loose glass as seen in Figure 8-3.

Figure 8-3: Offset Issue from Loose Glass Bed

The glass is encased between three walls: the front edge of the plastic base and the two sides. The back is held in place by the heated bed connector. I've found that placing a small wedge of thin material in the front edge and the side edge can lock the glass in tight. I just used sandpaper folded over on to itself and crammed into the edge. This was easy to do and the grip of the sandpaper kept it in place. The edge of the glass doesn't get that hot so there is no risk of it burning. I may want to remove the glass at some point so I don't want anything permanent in place.

You can see the paper in the picture below on the left side of the bed and in the front. I'm sure there are better repairs than sandpaper but it worked so well I left them in place. Thin metal shims would probably be a better option. I've even tested thin plastic prints as shims and they worked well.

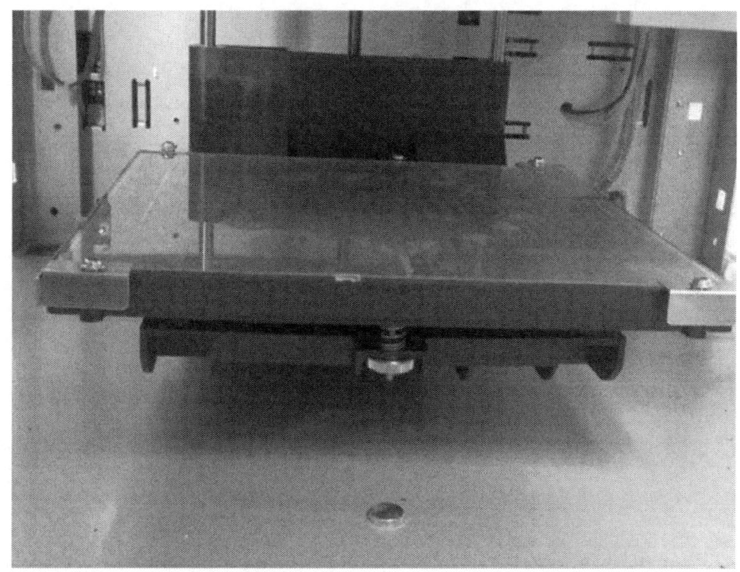

Figure 8-3: Sandpaper Inserts to Reduce Glass Movement

I also noticed that the bearing guide on the left side of the base is completely encased on all sides but the right side bearing is an open C shaped snap design. I found the lower bearing could shift as I moved the base from side to side. I jammed an old hobby knife blade in between the bearing and the plastic frame and that seemed to solidify it (Figure 8-4). Later I broke off a piece of the blade and used epoxy glue to hold it in place.

Figure 8-4: Bearing Shimmed with Exacto Knife Blade

Broken Bearing Mount

Some of the early Da Vinci models had a problem with the Y-axis bearing mounts cracking (Figure 8-5). This would cause wobble in the shafts that control the belts in the X and Y direction. If this is cracked you will get objects that print oblong or oval, not round.

I replaced one on my Da Vinci and I show how to do that on my YouTube Channel. Figure 8-1 shows the broken mount in my printer. It's covered under warranty if your printer is less than one year old but you may have to send it back. If you are not handy with fixing this, don't try it. I found it somewhat difficult to do for a beginner. I was never offered the option to send it back for repair. I was just sent a new bracket so I fixed it myself.

If your printer is still under its one-year warranty, contact XYZprinting via email at supportus@xyzprinting.com and let them know the issue. I suggest you take a picture of the failure and also check all four bearing mounts before contacting them. They will only replace what you show is broken. They sent me one replacement bracket. No extras.

Figure 8-5: Broken Bearing Mount

There are user created brackets on thingiverse.com but you need a good printer to make them before one cracks. This may be a good print to run once you are comfortable with the printer. Some people have access to metal CNC machines and have made metal replacements. This may be an upgrade you can purchase at some point.

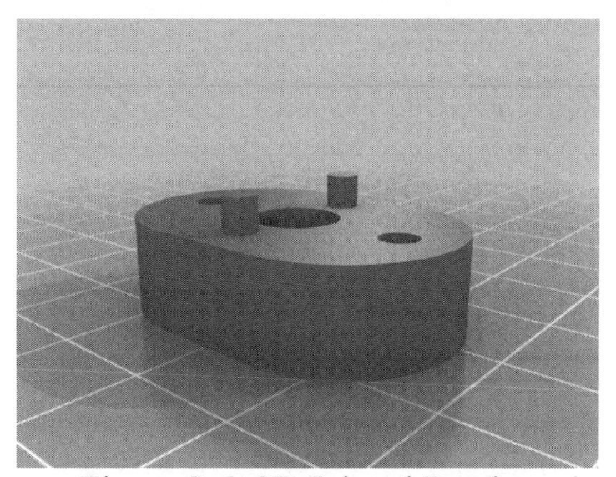
Figure 8-6: 3D Printed Bracket

Replacement Parts

The XYZ Printing website has a shop where you can buy some replacement parts. I hope to see it expand to include small pieces, like the bearing bracket, as well as the larger parts shown below. I've also seen broken Da Vinci 1.0 printers being sold for parts. This may be worth buying to have back-up components that are not available. I've had good results so far with the warranty support and very few issues so hopefully replacement parts won't be needed often but when I do need them, it's nice to know I can get them directly from the manufacturer.

Figure 8-7: Da Vinci Replacement Parts

Chapter 9 – Finishing Off Designs

A finished 3D printed design may not be completely finished to your liking. The ribbing or lines on the sides can create a somewhat rough or unfinished look on some designs. Of course this is from the layers of printed material. So can that be eliminated?
The answer is yes, if you want to put more finish work into it. Like a project made from wood, a plastic print can also be sanded smooth. This can help get rid of the rough edges. Another technique is using Acetone.

Acetone

You can get acetone in various forms and one popular form is in a nail polish remover. You want to use a 100% pure acetone remover though. Some polish removers say they contain acetone but looking at the bottle, you'll see it's diluted with other chemicals.

Using a cotton swab or cloth, you can spread a small amount of acetone on the 3D print and the outer surface of the plastic will begin to melt smooth. If you sand the print first followed by an acetone rub with a brush or cloth, the sides of the print can be smoothed out considerably.

Some people like to use a trick called acetone vapor. A small amount of acetone is placed in a metal container and then heated up to vaporize the acetone. The 3D print is then hung into the vapor for a very short time and the print will automatically begin to smooth out. I have tried this but I prefer the hand rubbing method for two reasons:

1) You can control the smoothness of the effect
2) Acetone vapor is flammable

Acetone is believed to have only slight toxicity so there should be no chronic health effects if normal precautions are used. By that, I mean use a well-ventilated area when you use acetone because high vapor concentration is not a good idea. Outside is the best location so there is plenty of fresh air around.

A big warning though is flammability. Acetone vapors are flammable. Though they have a high ignition initiation, you want to make sure the heated vapors have a open space to disperse.

Acetone as a Glue

Another finishing option to 3D printing is gluing several prints together to make a larger print. On the Da Vinci, you are limited to the 7.8" x 7.8" x 7.8" heated bed size. So if you want to make something larger than that, you can print it in sections and then glue the sections together to make a larger print. Acetone is perfect for this as well.

Because the acetone softens the outer surface of a 3D print, it is essentially re-melting the plastic so it can be bonded to another section of plastic. I use acetone as a glue to repair broken prints as well. Just a small brush of acetone of both pieces produces a molten surface that can then be pushed together. The acetone vapor will dissipate and the two plastics will harden together. So with this technique you can make some rather large 3D prints.

Acetone/Plastic Mix

Some people prefer to take small pieces of the plastic used in the print and grind them up into small flakes. Those flakes are then mixed with a small amount of acetone to make a liquidized plastic combination. The combination is then brushed on the design to fill in any low points or gaps. The acetone will dissipate and the plastic will be left behind and fused with the exiting print. This is actually a clever technique but requires trial and error to get the proper mixture. This also allows you to use the same color plastic to touch up, similar to the way you would use touch-up paint to cover a chip or scratch in your car door.

Carving, Grinding

Now if you don't want to fool with any chemicals like acetone then you can use normal wood working tools to finish a design. Once the ABS plastic is hardened it can be grinded on and carved at. I've used an Exacto knife to cut out sections and used a power belt sander to smooth out and reduce down sections of a large print. The only issue is heat. You cannot power sand it very long before the plastic begins to melt from the friction in the sanding belt. Quick little touches to the rotating belt work great.

Drilling

You can drill into the finished print as well but heat again is an issue. You need to keep the drill bit cold so running cold water over the drilled object helps a lot. You also have to drill through the layers not on the same plane as the layers. The layers of a 3D print can easily separate if there is enough force between layers and a drill will definitely produce that force.

Rock Tumbler

When I was a kid, rock tumblers were popular. You could find old rocks and then run them in a home rock tumbler and they would get polished shiny smooth by the banging around inside the rotating drum. You can do the same with small 3D prints. By placing small pebbles into a rock tumbler with the 3D print, the rotating drum will lightly bang the pebbles against the 3D print and smooth out the rough edges. This works best for items that don't have any defined edges or square shapes as it will try to round off every corner and you may not want that on some 3D prints.

Painting

You can paint 3D prints as well. After you get them sanded and smoothed out, use a plastic primer available at most hardware stores, and achieve the exact color you want. Of course you could just print in the color if it is available but that is not always an option.

Chapter 10 – Scanning, Dual Heads and Other Printers

3D printers continue to evolve. As I write this, XYZprinting has released the Da Vinci Duo 2.0. This is a dual print head 3D printer that can print two colors at once. This makes for a very interesting option. The printer is shown in Figure 9-1. You can see the dual catch boxes on each side below the extruders. The Da Vinci 1.0 has space for two plastic cartridges so it was planned early on to have one design that can do both single and dual print heads. The Duo 2.0 retails for $649.00 so it's 50% more expensive than the 1.0 but then you get two heads.

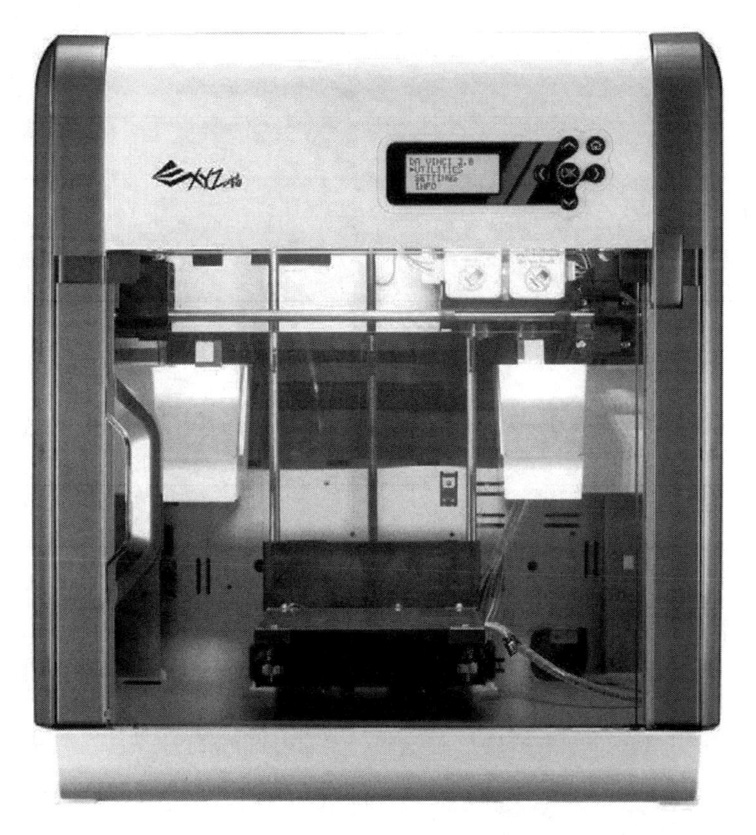

Figure 9-1: Da Vinci Duo 2.0 Dual Head 3D Printer

Since the designs are similar, everything I covered in this book should apply to the 2.0. Another printer XYZ is working on is one that has a built in scanner.

The Da Vinci 2.1 Duo AIO (All In One) Printer has a rotating base under the build platform and infrared scanners that will allow you to insert an object and then scan it so you can reproduce it on the 3D print platform. This was announced and shown at the 2014 Consumers Electronics Show (CES) so I don't have a lot of detail yet but this could really be a game changer.

There are scanners separate from the printer now but none built into the same unit. It's reported that the cost should be less than $1000 so that is incredible if it works as well as the Da Vinci 1.0 does. You can see a picture in Figure 9-2. (If you are reading this and you already own a Duo AIO then be aware that as I write this, I was dreaming of one day being in your position and hopefully I have one too.)

Figure 9-2: Da Vinci 2.1 Duo AIO 3D Printer/Scanner

Future Da Vinci Printers

At the 2015 CES, XYZprinting showed a future low cost PLA only printer that is expected to sell for $349.

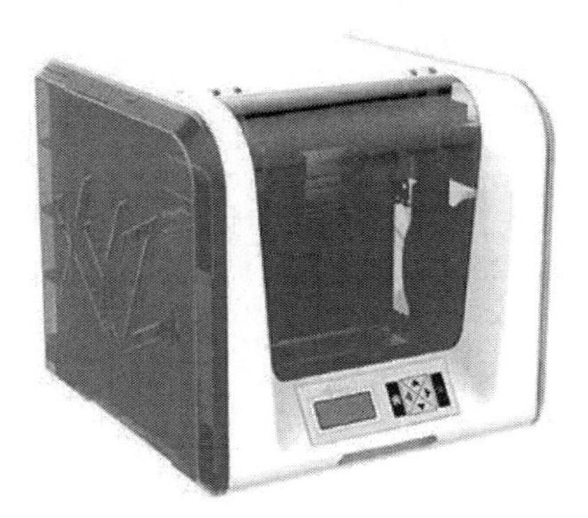

Figure 9-3: Da Vinci Jr

They also released a liquid resin printer that produces very high-resolution prints by using a laser to harden a liquid plastic. They call it the Nobel and it's suppose to sell in the $1500 range which is about half the cost of similar printers.

Frosting on the Cake

There are 3D printers that eject frosting instead of plastic so decorating cookies or cakes can be done to perfection. In fact at the 2015 CES, XYZprinting showed a food-printing machine they plan to release.

Future Developments

3D printing artificial body parts for medical students is already being done using special materials similar to human flesh. The sky is really the limit where this can all go but what you learn with a low cost Da Vinci 1.0 Printer and Tinkercad can launch you into a world of fun.

There are various types of filaments being released everyday. Flexible filament, filament with wood and metal particles mixed in so you can print wood and metal looking objects. Conductive plastic is also available for making cases that help shield the electronics inside from radio waves. Plastics that glow in the dark and all kinds of new colors are available. So it seems that home 3D printing is just getting started. It will be interesting to see where it is in five years.

Conclusion

I hope you found this book helpful and interesting and also educational as you learn how to 3D print the objects you will soon create. When I got mine I really didn't know if I would use it enough but before long it was so busy I had to get a second printer. And I paid for it by selling some of the creations I made. You can see them on my YouTube channel (https://www.youtube.com/user/beginnerelectronics) as I made many episodes around 3D printed projects. It is really a lot of fun and hopefully you have as much fun as I do.

You can always contact me with questions or comments via email at:

chuck@elproducts.com

I also have more information at my website:

www.elproducts.com

I also write books on electronics so that is a major part of my site. I also write a text blog there, which I try to keep active with 3D print articles and electronics related articles.

I also release weekly videos on 3D printing and other projects at my YouTube Channel:

www.youtube.com/beginnerelectronics

I actually have videos that show a lot of what is in this book so after you read a section, visit my YouTube Channel to see the topic covered in video.

Thanks for reading this book and I look forward to hearing what you created with your 3D printer!

26812268R00053

Made in the USA
Middletown, DE
07 December 2015